Noise & Health

Noise & Health

Thomas H. Fay, Ph.D., *Editor*

Columbia-Presbyterian Medical Center
New York, New York

The New York Academy of Medicine
New York, New York
1991

Associate Editor: Denis M. Cullinan
Cover designer: Marilyn Rose

The paper used in this publication meets the minimum requirements of American National Standard for Information Sciences—Permanence of Paper for Printed Library Materials, ANSI Z39.48-1984.

Library of Congress Cataloging-in-Publication Data

Noise & Health / Thomas H. Fay, editor.
p. cm.
Includes bibliographical references and index.
ISBN 0-924143-01-0 : $35.00
1. Noise—Health aspects. 2. Noise—Physiological effect.
3. Noise pollution. I. Fay, Thomas H., 1926- . II. Title: Noise and Health.
[DNLM: 1. Health. 2. Noise—adverse effects. WA 776 N7832]
RA772.N7N649 1991
363.7'4—dc20
DNLM/DLC
for Library of Congress 91-35009
CIP

Port City Press
Printed in the United States of America
ISBN 0-924143-01-0

Contents

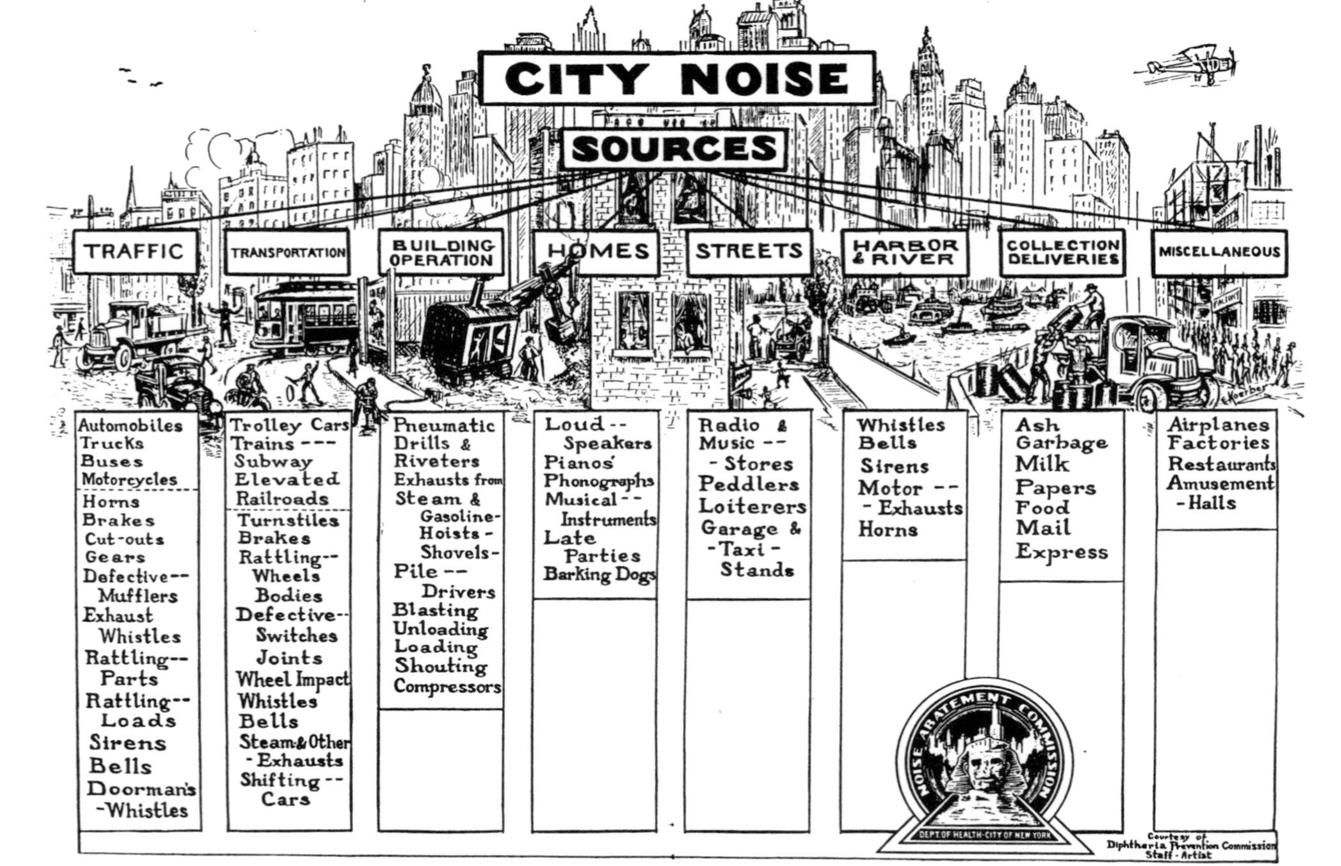

Endleaf for *City Noise: The Report of the Commission Appointed by Dr. Shirley W. Wynne, Commissioner of Health to Study Noise in New York City and to Develop Means of Abating It.* Edited by Edward F. Brown, E. B. Dennis, Jr., Jean Henry and G. Edward Pendray. Noise Abatement Commission, Department of Health, City of New York. 1930. (Courtesy of Diphtheria Prevention Commission Staff Artist.)

Foreword

Although it is generally recognized that sounds have profound effects on human behavior, only recently has attention been focused on the nonauditory effects of sound, especially those associated with health—the primary subject of this book.

Sounds provide many modes of communication. Complex linguistic and musical exchanges are the means of recognizing many physical and environmental phenomena. For early humankind, the identification of sounds was essential to survival. While this is to a large extent still true, in recent times well-being is more often linked to the suppression of sound.

We can simply categorize sounds as desirable or unwanted. The latter we usually describe as noise. According to Webster's, sound can be readily defined as "the sensation perceived through the sense of hearing . . . a tone . . . mechanical radiant energy that is transmitted by longitudinal pressure waves in a material medium (as air) and is the objective cause of hearing." *Noise,* on the other hand, is a much more ambiguous and subjective matter, and evokes negative emotions.

Sounds have an impact on many aspects of our lives, and whether highly organized or disparate, they evoke a broad range of emotions. A baby's cries, the din of a flock of starlings, or the powerful drama of unleashed thunder, while disturbing, are less unwelcome than the cacophony of urban life. City dwellers are plagued by a mechanical symphony of horns and sirens sounded by fleets of vehicles that echoes through concrete canyons while helicopters superimpose a staccato counterpoint from overhead. Escaping from such vexations by living on the upper levels of a high-rise apartment building is sometimes foiled by the seeming amplification of street sounds above these concrete corridors. Wearing individual listening devices represents a personal attempt to insulate one's self from these noises by masking them with a chosen and ordered sound of higher intensity. However, this expedient can dangerously isolate and distract the individual from the physical dangers in the immediate environment, and may cause hearing damage as well.

The human contribution to the world inventory of sounds has been an abundant one. While some sounds have been created intentionally, many are the unsought by-products of industrialization and urbanization. The toll taken by industry on the environment has been seriously addressed only in the last 20 years, and the problem of industrial noise has been one of the last of these to receive attention.

Reforms which began with child labor laws have evolved to highly-defined and broad-ranging worker protection codes promulgated by the Occupational Safety and Health Administration (OSHA). However, standards relating to hearing conservation are relatively new and narrow in scope, being concerned only with the direct effects of noise in the occupational setting on hearing.

The hazards posed by noise to human health have been aggravated by the shift from agrarian to urban ways of living. The urban setting provides a high population density and many varied sources of noise. We ourselves have become both sources and victims of environmental noise pollution. According to the American Conference of Governmental Industrial Hygienists, one-fifth of those in the United States aged 50-59 have experienced some degree of hearing loss from other than occupational exposures.

But noise may also cause nonauditory physical damage (to say nothing of psychological harm), and some individuals may be especially susceptible to noise or to certain of its characteristics. Intensity is only one of these characteristics. Unfortunately, no instruments have been devised that can assess noise in such a way that we can predict its unique effects on an individual human subject.

This book is a logical outgrowth of surveying the relationship of various environmental factors to health, and complements other workshops and studies conducted through the New York Academy of Medicine's Committee on Public Health. In order to form opinions, make health recommendations, and devise regulatory policies, currently available information about the effects of noise on human health must be collected, evaluated, and interpreted, and those areas that have been insufficiently studied must be delineated and the necessary data obtained. Given that nonauditory health effects can be substantiated and strategies for their control implemented, the risks to individuals and society must be balanced against other public health priorities and society's capacity to address them.

The study of the effect of noise on human health represents one of the last frontiers of sensory science, and we are indebted to Tom Fay and his colleagues for directing their efforts to this subject. We anticipate that this work will be not only a punctuation of what has been done in the past, but also a cornerstone of reference for what must be done now for the future.

EDWARD L. GERSHEY, PH.D
Chairman, Subcommittee on Environmental Health
New York Academy of Medicine

Preface

Noise & Health attempts to bring together in a single small volume basic information about the effects of noise on various aspects of health drawn from a wide variety of sources. Its main objective is to document that noise is a health hazard and to review what is known of its interactions with specific areas of health. The book addresses the question of the definition of noise, its sources, its physical and psychological effects, and remediation.

There is no single periodical that devotes itself to this admittedly broad area of subject matter. This is possibly due to the simple fact that noise is pervasive in most civilized societies and invades so many aspects of human health throughout our lifetimes. The cover artist has insightfully suggested this involvement through her use of the tentacled ampersand in the title, wherein noise and health appear to be entangled with one another. The literature covering the various aspects of this interaction is scattered across many different subject areas of worldwide scientific investigation, creating a daunting task for the reader who is interested in exploring them. It is essentially for this reason that the Committee on Public Health of the New York Academy of Medicine, through its Subcommittee on Environmental Health, decided to prepare a document that would address all of the major health effects of noise. The editor, as a member of this Subcommittee, was asked to assemble and serve as chair of the Working Group on Noise and Health, the members of which would serve as authors of the chapters to be included in the document. The group consists of eleven members representing different aspects of relevant professional expertise including audiology, cardiology, chronobiology, engineering, gastroenterology, immunology, neuroendocrinology, psychology, and speech-language pathology.

The task of assembling complete copies of significant world periodical literature for use by the authors was undertaken by the editor in conjunction with Dr. Jacqueline Messite with the assistance of the staff of the Office of Public Health and the Library of the New York Academy of Medicine. All available literature search lists were provided for the authors, who selected what they wanted to review. Almost all requests were obtained, albeit over a substantial period of time. Since the book is not intended to be a compendium, the authors' selections represent their own judgement of an accessible but representative sampling of recent world literature covering their topics, with additional selections by the editor.

The ways in which the various topics of this book are treated by the authors are largely dependent upon the availability and extent of published scientific research in the various areas, some of which are relatively rich in sources, while others suffer from a paucity of studies that address their focal interest. Hence sizable reference lists with many citations throughout are hallmarks of some chapters, while anecdotal, inferential, and informal writing is encountered in others.

We have tried not to entertain suppositions and surmises when properly controlled research findings are not available, but we have not shied away from calling attention to areas, both obvious and obscure, that need the bright and unbiased light of the scientific regimen shined upon them. We know of no deaths directly attributable to noise as the primary cause, but who among us would venture to say that it never plays a role in a complex of contributing factors that could eventually overwhelm an individual life? While hearing impairment has been the most obvious health effect of noise for centuries, it is by no means the only one, and herein is treated simply as one of a variety of effects. Following the section on the definition and sources of noise, its effects on the cardiovascular, neuroendocrine, immunological, and gastrointestinal systems as well as fetal development are considered. We then discuss susceptibility to noise-induced hearing loss and acoustic trauma, followed by surveys of the general effects of noise on sleep, the ear and hearing, learning, cognitive development, social behavior, attitudes, and community responses. There follows a discussion of remediation by way of specific strategies for noise abatement, a consideration of the economic effects of noise and its abatement, and a brief review of noise legislation. We conclude with a discussion of public awareness of the hazards of noise, a review of some needs for public education with reference to a school demonstration project, and a brief list of resources for limited materials and other assistance.

The low priority given to noise abatement by the Federal government has left the responsibility to state and local governments, which results in considerable variation in commitment across the various jurisdictions. The low priority also extends to sources of funding for research and demonstration projects. The Sponsored Program Information Network (SPIN) that searches for all funding sources in foundations and government agencies lists only one source of funding for noise abatement projects in the entire nation for 1991, and that is the National Institute of Environmental Health Sciences Extramural Program. That it happens to be a Federal funding source is a hopeful sign. Federal funds for other kinds of noise-related projects have been available on a limited basis. The printing of this book was made possible partially by the generous support of the National Institute for Occupational Safety and Health (NIOSH), of the Centers for Disease Control, United States Public Health Service, and we gratefully acknowledge that assistance.

The editor is indebted to each of the chapter authors for their freely-given efforts to this voluntary project. Their labors are deeply appreciated. Very special thanks are due to Jacqueline Messite, M.D., Executive Director of the Office of Public Health and Executive Secretary of the Committee on Public Health of the New York Academy of Medicine, for her staunch support and creative resourcefulness in achieving the production of this book. Thanks are also due to Ana A. Taras, Program Associate, of the Office of Public Health for her invaluable assistance, and to Joan Bonanno, Secretary, Office of Public Health, for assistance with the manuscripts. The editor is grateful for the richly professional editorial skills of Denis Cullinan along with his consistently generous helpfulness in so many ways. Special thanks are also due to the cover artist, Marilyn Rose, for her incisive designs. And to Dr. Annette Zaner for her cheerful and ready assistance with proofing the manuscripts. The editor expresses his deep appreciation to Josie Del Toro and Christina Fernandez, his secretaries at Columbia-Presbyterian Medical Center, for their assistance in preparing and disseminating the reading materials for the authors and for their help in preparation of the manuscripts.

Finally, the editor wishes to express his respect and appreciation to the late professors John W. DeBruyn, Grant Fairbanks, Joseph McV. Hunt, and Edmund P. Fowler, Jr., each of whom has contributed masterfully to the fight against noise in uniquely pioneering and preparatory ways, and to his late maternal grandmother, Mary Anderson Long, who demonstrated by dignified example just what serenity and quiet are all about.

THOMAS H. FAY
Columbia-Presbyterian Medical Center
New York, New York

Notes on Contributors

Barbara Ashkinaze, M.S., M.P.A. is an audiologist in Hearing and Speech Services, Department of Otolaryngology, The New York Hospital—Cornell Medical Center, New York, New York 10021.

Arline L. Bronzaft, Ph.D., is Professor of Psychology at Lehman College of the City University of New York, Bronx, New York 10468.

Thomas H. Fay, Ph.D., is Director of the Speech and Hearing Department at Columbia-Presbyterian Medical Center and Professor of Clinical Audiology and Speech-Language Pathology, Department of Otolaryngology, Columbia University College of Physicians and Surgeons, New York, New York 10032.

Edward L. Gershey, Ph.D., is Associate Professor and Director of Laboratory Safety at The Rockefeller University, New York, New York 10021.

Marc B. Kramer, Ph.D., is Director, Hearing and Speech Services, and Adjunct Assistant Professor of Otolaryngology in Audiology, Department of Otorhinolaryngology, The New York Hospital—Cornell Medical Center, New York, New York 10021.

Jane R. Madell, Ph.D., is Director of Audiology at the New York League for the Hard of Hearing, 71 West 23rd Street, New York, New York 10010.

Anita T. Pikus, M.A., C.C.C., is Chief, Clinical Audiology, National Institute on Deafness and Other Communicative Disorders, Bethesda, Maryland 20892.

Charles P. Pollak, M.D., is Director, Institute of Chronobiology, Sleep-Wake Disorders Center, New York Hospital—Cornell Medical Center, 21 Bloomingdale Road, White Plains, New York 10605; and Associate Professor of Neurology, Cornell University Medical College, New York, New York.

Lawrence W. Raymond, M.D., is Associate Medical Director of Exxon Company USA (P.O. Box 2180, Houston, Texas 77252) and Adjunct Professor of Medicine at the University of Texas School of Public Health.

Richard P. Sloan, Ph.D., is Associate Professor of Clinical Psychology (in Psychiatry), Columbia University, College of Physicians and Surgeons, New York, New York 10032.

Samuel Stempler, P.E., is Assistant Commissioner for the City of New York, Department of Environmental Protection, Bureau of Air Resources, 59-17 Junction Boulevard, Elmhurst, New York 11373-5107.

Annette Zaner, Ph.D., is Director, Communication Disorders Department, Mount Carmel Guild, Newark, New Jersey 07102; and Adjunct Associate Professor, Department of Speech, The City College of New York, CUNY, New York, New York.

Definition and Sources of Noise

ANNETTE ZANER

NOISE DEFINED

The simplest definition of *noise* is *unwanted sound.* And sound, whether wanted or unwanted, liked or disliked, and safe or hazardous, may be defined as a disturbance in the particles of air or any other medium surrounding or in contact with a vibrating source. If the medium is air, rapid changes in atmospheric pressure are propagated outward in all directions in alternating waves of compression and expansion until they encounter a reflecting or absorbing surface. If the surface happens to be a living creature's healthy ear, it will convert the pressure changes into patterns of nerve impulses that are transmitted to the brain, where they will be perceived as sounds that are unique to the specific patterns of the original vibrations. It is at this point that a conscious decision may be made as to whether a specific sound is wanted or unwanted, depending on the perceiver's attitudes toward it. Note that sound may contain sufficient intensity to cause actual physical damage to the ear receiving it, or be capable of inducing a stressful response, and still be considered a desirable auditory experience.

There is little question that noise includes those sounds of great intensity that emanate from heavy industrial machinery, from ground and air transportation vehicles, and from various home appliances. These are the unwanted sounds to which great numbers of individuals in our society are unavoidably exposed on a daily basis (Kryter 1985). These are the sounds that usually evoke responses of annoyance or extreme displeasure.

What of those sounds that may be troublesome only to some individuals, or to small groups of individuals? These are the sounds from neighbors, pets, occasionally-used household tools or machinery, and amplified music or speech emanating from commercial settings, all of which may intrude into one's environment. These noises are sometimes faint, but can cause significant discomfort or disturbance. A given sound may stimulate pleasure or displeasure, depending upon the attitude of the listener. Therefore, other than those of very high intensity, no one noise is likely to elicit the same response from different people (Burns 1973).

Kryter's (1985) definition of noise as "an audible acoustic energy that adversely affects the physiological or psychological well-being of people" would appear to include *all* unwanted sounds as noise. In what follows, we shall implicitly accept this definition as valid.

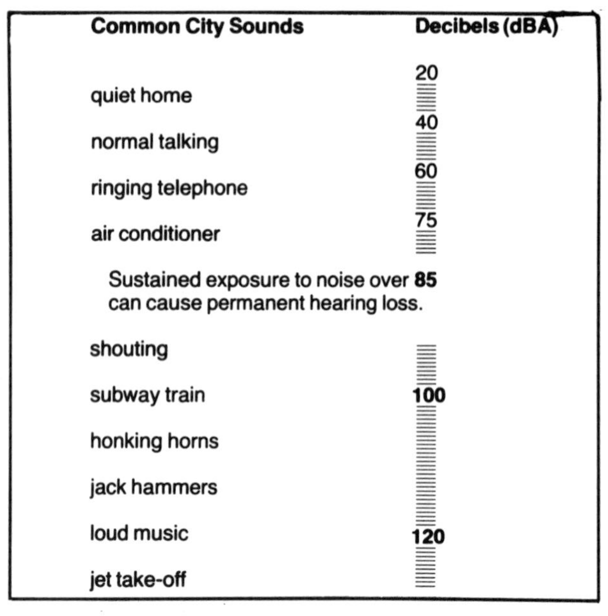

FIGURE 1. Intensity reference scale (With permission from the Council on the Environment of New York City.)

FIGURE 1 provides an intensity reference scale by which the levels of noise that will be mentioned farther on can be assessed.

SOURCES OF NOISE

Because sound is associated with every type of human activity, and because it is emitted from things that have moving parts (White 1975), sound is virtually everywhere.

Those sounds not welcomed by the hearer and sounds that have a known adverse effect on human beings have been present as environmental pollutants for thousands of years (Bragdon 1971; Pawlak 1978). The literature contains numerous examples of unwanted sounds. We can read accounts of noisy delivery wagons (the ancient counterpart of today's trucks) on the cobblestone streets of ancient Rome, Old Testament stories of excessively loud music, or tales of the annoying barking of dogs and squealing of pigs in old poetic allusions (Lipscomb 1974). In more recent history, from the end of the 19th through the beginning of the 20th centuries, noise has proliferated as a result of the Industrial Revolution and the accompanying increase in urbanization. With these events came a concomitant increase in the demand for transportation and, inevitably, an eager market for powered vehicles and other products (Crocker 1984).

The growth and utilization of noise-producing and noise-related technology in modern civilization are proceeding at such an accelerated rate that it is practically impossible to compile a catalogue of noise sources that will not quickly become out-dated. Nevertheless, we shall attempt to do so, with the qualification that the following is by no means to be viewed as an exhaustive list.

TRANSPORTATION NOISE

Survey data collected in the early 1970s indicate that, with respect to noise levels of intermediate intensity, urban traffic provides the most significant source of annoyance, followed by airport and construction activities. As far as more intense noise levels are concerned, highway noise and aircraft noise were cited as most bothersome. According to the U.S. Environmental Protection Agency, "Over 40 million residents of the United States . . . [were reported to be] disturbed by urban traffic noise and some 14 million by airplane traffic noise." (National Academy of Sciences 1977).

Surface Transportation

Road vehicles are an increasingly troublesome source of noise "everywhere growing in intensity, spreading to areas until now unaffected, reaching even further into the night hours" (Bugliarello et al. 1976).

Automobiles. The noise heard inside a fast-moving American automobile is mainly produced by the contact of the tires with the road. It is transmitted to the interior of the car through body mounts, steering controls, and its suspension system. In European and Japanese cars, the noise is generated chiefly by engine parts and the exhaust systems (Bugliarello et al. 1976). Generally speaking, more costly automobiles have lower interior noise levels than do economy models or sports cars. However, this is true only if the car windows are kept closed. If the windows are open, the noise level inside the car will be as great as that of the surrounding traffic.

Buses and Trucks. Highway as well as urban street noise problems are exacerbated considerably by diesel-powered buses and trucks. In diesel engines, ignition takes place at higher pressure than in gasoline-driven engines. Furthermore, the diesel engine emits more noticeable airborne vibration, especially when the vehicle is heavily loaded, when it is driving on an upgrade, or when it is accelerating from a stop. For these reasons, and because they usually operate at full speed and maximum power, diesel-powered vehicles produce much more highway noise (Bugliarello et al. 1976). In urban settings, public transportation systems consist primarily of diesel-powered buses, which are especially noisy.

Motorcycles. Motorcycles are capable of producing very high levels of noise that are particularly disturbing. Not even the riders themselves can always be effectively shielded from this noise. At 50 feet (comparable to the distance from the roadway to a home in a suburban neighborhood), older-model motorcycles and those with poor or defective mufflers can create noise loud enough, given sufficient exposure time, to cause hearing impairment. Racing models and those equipped with muffler cut-out devices can be even louder. Although motor scooters are generally less noisy than motorcycles, the noise that scooters create is often equally disturbing because it has relatively more energy at the higher frequencies of the spectrum (Bugliarello et al. 1976).

Other Roadway Factors. In urban centers where there are a large number of passenger cars, heavy commercial vehicles, buses, motorcycles, and a variety of other surface transportation, the traffic congestion produces driving patterns that include frequent rapid acceleration. The number, type, and weight of the vehicles plus the rapid acceleration factor cause noise peaks which are superimposed on an already excessive ambient noise level (Bugliarello et al. 1976).

At roadsides, along highways, and in urban areas close to interstate highways, transportation noise is caused by a number of other factors. Among them are road surface (rough pavement tends to increase tire noise), gradients, width, and configuration. If the road is recessed, additional noise may be created by reverberation against the road walls. In a given time period, noise will increase with the number of vehicles on a road. Since heavier vehicles make more noise, there will be an increase in noise with an increase in the ratio of heavier to lighter vehicles. As speeds increase, however, this latter relationship becomes less pronounced. Therefore, the composition of traffic influences noise on highways to a lesser degree than it does on urban roads (Burns 1973).

Rail Vehicles. A primary source of noise in rail vehicles such as subways, elevated trains, and other municipal rapid transit, is the interaction between the wheels and the guideway. Noise levels are often excessively high when the equipment is old and maintenance practices are deficient (Bugliarello et al. 1976). Older subway systems, such as those in New York City, tend to have the highest interior noise levels. But even in newer systems, with designs aimed at noise reduction, the noise levels are very high. When a rail transit vehicle goes through a tunnel (the most common circumstance in urban systems), the noise level in the interior of the vehicle will increase greatly (Bugliarello et al. 1976).

In subway stations, the noise is often unbearably loud. Screeching and other noises produced by wheels against rails (especially when the train is rounding a curve), and the positive resonance and reverberative effects created by the tunnels, combine to produce an incredible amount of noise which represents an enormous disturbance to passengers waiting on subway platforms. The disturbance is particularly great for those passengers waiting on platforms at local stations where no noise-absorbing barriers have been installed, and where express trains with many cars go by at great speeds.

For railroads that travel long distances, the basic sources of noise are locomotives, wheels, coupler interaction, various structural vibrations, and refrigerator-car cooling-system motors (EPA 1974). In all urban and suburban settings in modern developed countries, the noise from surface transportation is now so pervasive—whether inside or outside the vehicle—that almost no one can escape this source of annoyance.

Air Transportation

For those people living close to either civilian or military airports, aircraft noise clearly causes great disturbance. It can be far more irritating than noise arising from surface transportation.

As a result of an enormous increase in the utilization of air transportation following World War II, the noise created by airplanes has become a major problem. An escalation of the noise level attributable to aircraft occurred toward the end of the 1950s, at which time jet airplanes came into widespread use. A further escalation in aircraft noise is related to the commercial operation of supersonic transports.

From the very beginning of aviation all aircraft have been noisy. "All aircraft engines are heat engines that convert rapidly expanding gases into thrust" (White 1975), and it takes only a small proportion of waste energy, converted into audible sound waves, to cause a great nuisance.

Propeller Aircraft. Although propeller aircraft currently make up only a small part of our commercial air fleets, they are still used to a large extent for private, business, corporate, and instructional purposes (White 1975). The noise emanating from propeller planes comes from two sources: the noise created by the rotation of the propeller as it periodically disturbs the air (mostly low frequency noise) and the noise from the unmuffled engine exhaust.

Helicopters. Helicopter noise sources include a combination of (1) blade slap, (2) engine exhaust, (3) tail rotor rotation, (4) main rotor rotation, (5) the gear box, and (6) the turbine engine (White 1975). The result of the interaction of the noises produced by these sources is "a distinctive low-frequency throbbing sound that propagates over great distances" (White 1975). Particularly bothersome high noise impacts are generated by extended low-altitude operation, and are aggravated by helicopter hovering, such as is common in traffic reporting and police operations. These activities have become more and more frequent.

Helicopters are especially well-suited to provide transportation to and from city centers. Because of the high level of radiated noise that they produce, public acceptance of these aircraft has been largely limited to such short-hop functions. However, with the increase in their utilization and the close proximity of landing pads to high-density, high-rise dwellings, helicopters are becoming an ever greater annoyance to those who live and work near heliports.

Jets. Most large passenger aircraft are now powered by jet engines. Jet aircraft produce two kinds of engine noise, "a low-frequency 'roar' caused by the mixing of hot exhaust gases with . . . air around the aircraft . . . [and] . . . a high-frequency 'whine,' generated in the compressor section of the engine" (Bugliarello et al. 1976). With variation in the type of aircraft and the power plant there is variation in the composition and level of noise.

There are basically two types of jet engines: turbojets and turbofans. Turbojets were the first generation of jet engines used in commercial service. The noise sources include, primarily, the turbulent mixing that occurs along the boundary between the high-velocity jet exhaust and the stationary atmosphere. Other sources include the compressor and turbine, and the unsteady combustion (Crocker 1978).

During the 1960s the first turbofans (also known as fan-jets or bypass jets) were used on passenger aircraft. These were used instead of turbojets because they proved to be more efficient. They also happened to be quieter—on takeoff. Approach or landing noise was not reduced (Crocker 1984). The turbofan engine is capable of quieter operation because it incorporates a thrust-producing fan, which, mounted outside and behind the inner engine case, effectively blocks a portion of the engine noise, and because it produces a lower jet-exhaust velocity than does the turbojet engine (Raney and Cawthorn 1979).

Supersonic Transports (SSTs). Because of our ever-increasing desire to travel long distances in shorter time spans, SSTs have emerged as most appealing transportation vehicles. When an airplane flies faster than the speed of sound, however, an aerodynamic phenomenon known as a "sonic boom" occurs. An audible shock wave is created by the rapid compression of air, and is pushed ahead of the plane, forming a cone of increased air pressure, with the plane at the tip of the cone (Bugliarello et al. 1976). The boom can be audible over an area of 50 or 60 miles. Takeoff noise, which is generally most intense for all aircraft, is even greater for SSTs, which require greater takeoff speeds.

As with all other sources of noise, airplanes are a nuisance not only to those individuals who are involved in their operation (passengers and personnel), but to those who are involuntarily exposed as well. Most significant about aircraft noise, however, is that the segment of the environment and its involuntarily exposed population is much greater in relation to the segment containing those who are directly involved in operations (Bugliarello et al. 1976).

Water Transportation

Ships. Today's ships are usually equipped with powerful engines that produce high noise levels. The engine noise is typically combined with noise from air conditioning, ventilation systems, and auxiliary electric generators, all of which tend to be transmitted and reinforced by the very structure of the ship. This cacophony subjects passengers and crew alike to excessive noise levels for long periods of time (Williams 1985).

Hovercraft. Hovercraft are a relatively new type of transportation utilized mainly in Europe, and especially in Britain. (They are, however, gaining in popularity in the United States, and are in limited use as river transportation in New York City.) Their chief use is as high-speed ferries, crossing rivers and channels. The noise produced by hovercraft, which can be extremely loud, emanates primarily from massive centrifugal lift fans that support the ship on a cushion of air, the driving propellers, and the engines. The noise levels of earlier versions are comparable to those of single engine aircraft (White 1975), while newer designs are slightly quieter.

Because the larger passenger hovercraft are relatively few in number, because they usually do not operate in heavily populated areas, and because their journeys are of relatively short duration, the noise mainly affects the passengers and crew (Williams 1985). As they become more common on metropolitan waterways, their noise pollution will affect even larger segments of the population.

Hydrofoils. Hydrofoils are high-speed craft that literally rise up and skim across the surface of the water on water-ski-like runners. They are thrust forward by propellers that produce the same kind of noise as the airborne versions, except that these remain on the surface and create considerable disturbance as they pass occupied areas.

Recreational Vehicles

Snowmobiles. The sources of snowmobile noise, which is considerably greater in older than in newer models, are the exhaust, the carburetor air inlet, and the engine (Crépeau 1978). Because of their inherent noisiness, legislation has been designed to control their noise output in order to protect both the operator and the passenger, and to reduce the environmental impact. Customer demand for high horsepower, however, tends to work against attempts at noise abatement (White 1975).

Motorboats. The main noise sources for motorboats, whether those with inboard or outboard engines, are the exhaust, the engine, and the carburetor air intake (Crépeau 1978). Although noise from most modern motorboats does not generally present a serious noise hazard to the operator (White 1975), its impact on others is often great. An exception is the recent appearance of high-powered speedboats ("cigarette boats") that produce unmuffled, high-intensity sound levels capable of inflicting permanent hearing damage on their occupants. They are extraordinarily annoying environmental polluters as they speed along waterways adjacent to populated areas.

Hydroplanes. Propellers driven by gasoline engines force these flat-bottomed watercraft at high speeds over the surface of otherwise unnavigable bodies of water. With noise levels equivalent to those of aircraft and hydrofoils, they regularly penetrate the natural quiet of such wilderness areas as Florida's Everglades and other marshy wetlands.

Other Recreational Vehicles. Noisy gasoline engines are also used to power a variety of other vehicles used for sport that affect both urban areas and the once-pristine countryside. Landrovers, dirt bikes, and even powered skateboards, water scooters, skis, and surfboards are now in widespread use. Perhaps the greatest threat to natural quiet associated with the inaccessibility of a location is the heralded mass production of the smaller recreational versions of hovercraft that are presently available. These fan-lifted propeller-driven craft can literally glide over land, water, ice, deep snow, and marshy wetlands on a cushion of air without touching the surface. Capable of carrying two to six passengers, these vehicles, with noise levels in excess of 85 dBA, will intrude into areas heretofore inaccessible to other vehicles.

INDUSTRIAL NOISE

The potential ill effects of industrial noise first became a serious concern as a result of the Industrial Revolution, with its proliferation of machinery that often involved the hammering and grinding of metal against metal. Machines had been in use considerably before that time, however, with increasingly deleterious noise effects. Some simple machines (such as screws and wheels) were used as early as the third and second centuries B.C.; water and wind wheels were used in the tenth century A.D.; and by the eleventh and twelfth centuries, water-driven forge hammers and forge bellows came into use. In the thirteenth and fourteenth centuries, mills were used for pulping, sawing, grinding, metal-stamping and wire-drawing. Eventually, certain occupations (such as blacksmithing and boilermaking) became synonymous with disabling hearing loss (Crocker 1984).

TABLE I. Industrial Activities and Associated Noise Sources

Activity	Noise Source
Product fabrication Metals (e.g., boilers, cans)	Metal stamping and riveting
Molding: plastics and glass bottles	Pneumatic control devices and turbulent mixture of high-pressure air
Product assembly (e.g., automobiles, aircraft, household appliances)	Impact wrenches Riveting Grinding
Power generation (thermal and nuclear plants)	Turbine generators Air compressors
Large-scale processing (oil refineries and steel plants)	Broadband noise sources include furnaces, heat exchangers, pumps, compressors, and air and steam leaks

Paralleling the increase in the intensity of noise due to the expanding use of road vehicles in transportation, industrial noise has grown in both intensity and impact as a result of increased construction, widespread mechanization of agricultural methods, expanding utilization of powered household appliances, and the continued spiraling of the development, manufacture, and marketing of the items and objects involved in this progression (Bugliarello et al. 1976).

Industrial noise is created both indoors, for example, in plants for metal, glass, or plastic product fabrication and molding; product assembly; and power generation and processing; and outdoors, as in construction activity utilizing equipment for earthmoving and materials handling, as well as stationary, impact, and other types of equipment.

Indoor Industry—Plants

Indoor industrial noise is experienced both internally, i.e., within the plant itself, by those employed or otherwise occupied there, and externally, i.e., outside the plant, by those who live or work in or happen to pass through the surrounding areas.

Internal Noise. Bugliarello et al. (1976) presented a summary of a study reported by the EPA (December, 1971) of internal industrial noise as it relates to the various previously mentioned activities

TABLE II. Specific Noise Sources and Associated Noise Levels

Noise Levels in dBA	Noise Source
130+	Turbine or jet engine testing
120-129	Riveting
119-120	Metal-forming machinery
100-109	Wood forming (sawing, planing, lathe operations)
	Textile looms
	Large printing presses
	Paper manufacture
	Boiler room operations
	Plastic- and rubber-molding machinery
	Tumble cleaners
	Punch presses
90-99	Food canning
	Metal shears
	Steel product fabrication
80-90	Textile dyeing
	Apparel manufacture

within industrial plants. Eighty percent of the noise levels were shown to be above 80 dBA, and 20 percent above 95 dBA. Some of the items from their summary appear in TABLE I.

TABLE II furnishes a brief list of some of the specific sources and noise levels cited by Bugliarello et al. (1976).

External Noise. Industrial noise is typically "transmitted by an industrial plant to the outside . . . [via] . . . roof ventilators, open windows, steam injectors, compressors, and diesel engines" (Bugliarello et al. 1976). Industrial plants were originally built close to residential areas in order to insure easy access for laborers. Such sites still exist and often consist of a multiplicity of noisy plants. When experienced together, the noise from the different plants impacts greatly on the surrounding community. Because of vast improvement in the availability of transportation, however, more recently developed industrial sites have been located away from residential areas. The negative impact on the environment of the noise from the newer, more remote plants is therefore considerably less (Cunniff 1977).

OUTDOOR INDUSTRY—CONSTRUCTION

Although noise generated by various pieces of construction equipment is of very high intensity, it is generally not as great as the noise produced by industrial processes indoors. An exception is the noise levels of impact pile drivers, which reach 105 dB at 50 feet

TABLE III. Construction Equipment and Associated Noise Levels

Noise Source	Noise Levels in dBA at 50 Feet
Earthmoving equipment	
Compactor	71-75
Front loader	71-85
Backhoe	71-94
Tractor	77-96
Scraper or grader	80-97
Paver	86-90
Truck	83-95
Materials handling equipment	
Concrete mixer	74-89
Concrete pump	80-85
Crane (movable)	76-87
Crane (derrick)	87-91
Stationary equipment	
Pneumatic wrench	83-88
Jackhammer and rock drill	82-98
Impact pile driver	95-105 (peaks)
Other equipment	
Vibrator	63-83
Saw	72-82

noise levels of impact pile drivers, which reach 105 dB at 50 feet (Bugliarello et al. 1976). As Cunniff (1977) points out, however, "Clusters of equipment at construction sites can produce a steady roar from early morning to the evening hours . . . for relatively long periods of time."

A sampling of noisy construction equipment with their associated noise levels is listed in TABLE III (EPA 1971).

The workers who are at greatest risk of harm from industrial noise are those closest to the actual sources, whether these are inside plants or at outdoor construction sites. The noise sources associated with industrial as well as office equipment are so pervasive today that sound emanating from them has become difficult to escape.

NOISE IN THE HOME

One's home, thought to be a "sanctuary from damaging noise" (Bugliarello et al. 1976), is nevertheless an environment that houses a multitude of potentially noise-producing items. However, the levels of noise created by these sources, as well as the effect of the noise on the residents, depend on a number of factors, including the location of the home itself and its unique acoustic properties. For in-

TABLE IV. Typical Noise Levels of Home Appliances

Appliance	Noise Levels in dBA
Food mixer or processor	62-89
Whistling kettle	81
Pop-up toaster	78
Coffee percolator	54
Washing machine	49-72
Dishwasher	55-71
Garbage disposal	67-93
Coffee grinder	75-80
Alarm clock	64-76
Vacuum cleaner	60-85
Hair drier	59-71
Electric toothbrush	49-60
Electric shaver	48-80
Flush toilet	76-82
Air conditioner	50-69
Doorbell	79
Telephone	77
Sewing machine	70-75
Electric drill	91
Electric lawn mower	80-90

gested, noisy urban street? Are its rooms acoustically live or acoustically dead? A particular kitchen appliance, for example, may be perceived as noisier in a country home as opposed to a city home, because the higher ambient noise level in the urban setting tends to mask the appliance's noise. Furthermore, the noise produced by that same appliance will be louder if the room in which it is operated is live, having hard walls, ceiling, and floor that add a reverberation component to the noise emanating directly from the source. Compounding the problem further, noise from some household sources is structure-borne in addition to being airborne. Noise and vibration from washing machines, loudspeakers, vacuum cleaners, and other sources that make direct contact with walls, floors, and ceilings will be conducted by the structure itself, thereby causing great annoyance to residents in adjoining dwelling spaces as well as to those in whose home the noise is actually being generated (Jackson and Leventhall 1985).

Noise in the home is usually made up of one or more of the following: human-generated noise (voices, objects dropped, walking, jumping), appliance noise (indoor and outdoor motor-driven appliances), building equipment noise, and background noise entering from the outside (Bugliarello et al. 1976). A partial list of home appliances and their typical noise levels appears in TABLE IV (Jackson and Leventhall 1985; EPA 1971).

A recent addition to our ever-growing list of labor-saving devices, the leaf blower, has found its way from suburban to urban use. These gasoline motor-driven blowers are carried like back-packs by the operator, who directs tubular, hand-held, air-emitting nozzles before him, thus blowing leaves and other debris from paths. They have been in use by the New York City Parks Department to provide a presumably efficient means of moving leaves and refuse, and even for returning sand to designated play areas. The author measured the noise levels of a leaf blower in use at the Bleecker Playground in Manhattan from a location 3 feet inside an open window on the second floor of a building facing the playground. At this distance of approximately 200 feet from the operator the levels were 70 to 72 dBA. When measured outside at a location approximately 75 feet from the operator, where the ambient levels normally ranged from 58 to 62 dBA, blower levels were 82 to 85 dBA.

As our quest for greater ease, efficiency, and speed in completing our household tasks continues, and as technological development progresses, the number of noise sources and the level of noise in the home are destined to increase.

EMERGENCY SIGNAL NOISE

Sounds that are purposely produced in order to alert people to an emergency need by definition to be very loud, attention-getting audible signals. Sources that fall into this category are police, fire, ambulance, and rescue-vehicle sirens. In New York City, noise from such sirens has been measured as high as 124 dBA at three feet (Bugliarello et al. 1976).

Another category of signal noise includes sounds purposely produced with the intention to alert, arouse, interrupt, or otherwise call the attention of individuals to impending danger or to acts of anti-social behavior. These signals include noises emanating from burglar alarms, smoke and fire alarms, auto-theft alarm systems, automobile horns, and vehicle back-up signalling devices.

Although all such signal devices can be considered to have a socially beneficial purpose, the sounds they produce are too often experienced as noise by the individuals who are not necessarily directly affected by the potential emergency or danger that they signal.

OTHER SOURCES

Virtually endless numbers of sounds arise from sources in addition to those thus far noted that are or can be experienced as noise. Some of these are: (1) television sets; (2) radio and hi-fi systems outdoors and in homes and vehicles; (3) musical instruments; (4)

disco music; (5) garbage collection; (6) public address systems; (7) pets; (8) playgrounds and children's toys; (9) ice-cream trucks; (10) bells; (11) noises associated with military activities; (12) gunshots; (13) combinations of community noise; (14) high-speed auto racing.

It is important to note that, although loudness is a major factor in experiencing sound as noise, other features of sound, such as constant or changing patterns of rhythm or stress, can cause a relatively quiet auditory signal to be experienced as noise. For example, in a library-type setting, even the whispered conversation of two individuals can become a most distracting noise to others; a high-pitched sound, no matter how quiet (such as the screech of chalk on a blackboard), can be experienced as a most objectionable noise; and an automobile horn, when honked more than once or twice, is frequently perceived as a nuisance.

REFERENCES

Bragdon, C. R. 1971. *Noise Pollution: The Unquiet Crisis.* Philadelphia: University of Pennsylvania Press.

Bugliarello, G., Alexandre, A., Barnes, J., and Wakestein, C. 1976. *The Impact of Noise Pollution: A Socio-Technological Introduction.* New York: Pergamon Press, Inc.

Burns, W. *Noise and Man.* 2d ed. 1973. Philadelphia: J. B. Lippincott Co.

Crépeau, G. 1978. Noise of transportation to travelers. In *Handbook of Noise Assessment,* ed. D. May. New York: Van Nostrand Reinhold Co.

Crocker, M. 1978. Noise of air transportation to nontravelers. In *Handbook of Noise Assessment,* ed. D. May. New York: Van Nostrand Reinhold Co.

Crocker, M., ed. 1984. *Noise Control.* New York: Van Nostrand Reinhold Co.

Cunniff, P. F. 1977. *Environmental Noise Pollution.* New York: John Wiley & Sons.

EPA (U.S. Environmental Protection Agency) December, 1971. *Report to The President and Congress on Noise.* Washington, D.C.

EPA (U.S. Environmental Protection Agency) July, 1974. *Proposed Emission Standards for Interstate Rail Carrier Noise.* The Bureau of National Affairs, Washington, D.C.

Jackson, G. M., and Leventhall, H. G. 1985. Noise in the home. In *The Noise Handbook,* ed. W. Tempest. Orlando: Academic Press.

Kryter, K. 1985. *The Effects of Noise on Man.* 2d ed. Orlando: Academic Press.

Lipscomb, D. M. 1974. *Noise: The Unwanted Sounds.* Chicago: Nelson-Hall Co.

National Academy of Sciences. 1977. Noise abatement: policy alternatives for transportation. In *Analytical Studies for the U.S. Environmental Protection Agency.* Vol. 8.

Pawlak, J. 1978. Preparation of noise control legislation. In *Noise Control: Handbook of Principles and Practices,* eds. D. M. Lipscomb and A. Taylor. New York: Van Nostrand Reinhold Co.

Raney, J.P., and Cawthorn, J.M. 1979. Aircraft noise. In *Handbook of Noise Control,* 2d ed., ed. C.M. Harris. New York: McGraw-Home.

White, F. A. 1975. *Our Acoustic Environment.* New York: Wiley & Sons.

Williams, D. 1985. Noise in transportation. In *The Noise Handbook,* ed. W. Tempest. Orlando: Academic Press.

Cardiovascular Effects of Noise

RICHARD P. SLOAN

Concerns about the non-auditory health effects of noise have been raised over the past several decades. Such early reports as those of Smith and Laird (1930), Laird (1932), Medoff and Bongiovanni (1945), and Abbey-Wickrama et al. (1969) have suggested that noise has various detrimental psychological and physiological effects. However, many of these early reports have been criticized for methodological inadequacies that serve to render their conclusions suspect.

The focus of this review is confined to the examination of research on the cardiovascular effects of noise. Specifically, the relationship between noise and coronary artery disease, as well as essential hypertension, will be examined. Only studies appearing in peer-reviewed publications have been considered.

In general, the effects of noise on health have been studied in two different ways: (1) field investigations and (2) laboratory studies. In the former, health data are gathered from existing populations exposed to varying degrees of noise. In the latter, subjects are studied in the laboratory during controlled exposure to noise. Studies in this category include investigations of both humans and animals.

Field studies of the cardiovascular effects of noise are illuminating in that they help to identify possible causal relationships between environmental events and health outcomes. The greatest asset of such studies is their ecological validity, i.e., the fact that they tell us something about real-world relationships. Rarely, however, are such studies, in and of themselves, sufficient to provide unambiguous causal inferences since their very strength, their relevance to the real world, is also their most serious drawback. In field studies, populations under study may differ in many ways other than by exposure to the putative causal environmental agent and for this reason there is no certainty that these other differences are not causally related to the health outcomes under investigation.

Laboratory studies contrast with field studies in that groups of subjects are exposed to different conditions which are assigned at random. Studies of this sort are considerably more precise because of the greater control of events that the laboratory permits. For precisely this reason, however, laboratory studies may be deficient in ecological validity. The extent to which their findings can be general-the real world is always questionable, and researchers constantly confront the need to make their laboratory work relevant to real-life situations.

This distinction is highlighted by the difference between the outcome variables of field and laboratory studies. In the former, investigators are concerned with the existence of cardiovascular disease itself, for instance, coronary heart disease, hypertension, and so forth. It is widely recognized that heart disease is a chronic condition, one which develops over many years, if not decades. Laboratory studies, on the other hand, cannot examine the development of actual disease, for both practical as well as ethical reasons. Accordingly, laboratory-based studies are restricted to the examination of the influence of noise on hemodynamic effects thought to be characteristics of the development of subsequent disease, such as increases in blood pressure and heart rate, and in levels of blood cholesterol or other blood lipids. Alternatively, laboratory studies may rely on the use of nonhuman subjects, in which the development of actual disease may be studied. In either case, actual heart disease in humans is not studied.

In the early 1980s, several thorough reviews of the nonauditory effects of noise have appeared (Thompson 1981; Cohen and Weinstein 1981; DeJoy 1984). These reviews suggest that although considerable research has been directed to the examination of this problem, major questions remain unanswered. They also agree that studies of the nonauditory effects of noise may depend upon its context and that it is overly simplistic to consider only the physical properties of noise in studies of its nonauditory health effects.

This review will focus on more recent studies in the literature, specifically those which have emerged in the 1980s. Both field and laboratory studies will be examined.

Laboratory Studies

Studies of Human Subjects. Andrén and colleagues (1980, 1982) have reported that brief exposure to loud noise stimulation produces increases in diastolic blood pressure in normal subjects, and that in hypertensive patients this increase was associated with peripheral vasoconstriction (1983). However, in this lattermost study, blockade of the vasoconstrictive response resulted in an increase in cardiac output, which has the effect of sustaining the elevation of blood pressure. Combined alpha- and beta-blockade abolished the increase in systolic blood pressure (SBP) and total peripheral resistance but, nevertheless, diastolic blood pressure (DBP) and mean arterial pressure (MAP) increased. Because the blood pressure elevations occur even after alpha- and beta-blockade, these investigators speculate that the hemodynamic response to exposure to loud noise involves central mechanisms leading to temporary resetting of the baroreceptors.

Eggertsen, Andrén, and Hansson (1984) demonstrated that in hypertensive patients, exposure to 10 minutes of 105 dBA noise produced significant increases in diastolic blood pressure, mean arterial pressure, and vascular resistance in the calf (but not in the forearm) and that only the last-mentioned effect was reversed by administration of a combined precapillary vasodilator and non-selective beta-blocker. Eggertsen et al. (1987) found that 30 minutes' exposure to the same level of noise produced significant increases in systolic, diastolic, and mean arterial blood pressure in subjects with mild essential hypertension. Both total peripheral and forearm vascular resistance increased during the noise exposure, while cardiac output and forearm blood flow decreased significantly. Both studies suggest that the noise-induced elevation in blood pressure was produced by vasoconstriction.

Colletti and Fiorino (1987) examined the effect of 30-second bursts of 100 dBA broad-band noise on hemodynamic response (HR) in ten patients with unstable angina. Unlike the Eggertsen studies, this experiment revealed that only brief increases in HR and left ventricular systolic pressure occurred. Exposure to this noise did not alter coronary blood flow or ECG morphology. However, the authors point out that premedication of patients with diazepam might have dampened any hemodynamic effects.

Using normal subjects, Ray, Brady, and Emurian (1984) reported that during a 30-minute performance task simulating one that might be found in a work setting, exposure to 10 minutes of intermittent pink noise (93 dBA) produced hemodynamic effects that did not diminish by habituation over three experimental sessions on consecutive days. Specifically, while noise did not influence heart rate or respiratory rate, it was associated with increases in mean arterial pressure and digital pulse amplitude over and above those produced by the task alone.

In another study of the impact of noise on cardiovascular outcomes during task performance, Linden (1987) exposed four groups of subjects to five minutes of no noise or one of three noise conditions (steady-state "real-life noise," variable "real-life noise," and 90 dBA steady-state white noise) while they performed a mental-arithmetic task. The groups did not differ in their blood-pressure responses, but heart rate reactivity in the group exposed to real-life noise was significantly greater than in the other three groups.

Recently, Carter and Beh (1989) have reported the results of a sophisticated laboratory study of the effects of 55 minutes of intermittent noise (92 dBA) on cardiovascular functioning during performance of a vigilance task. Diastolic blood pressure was the variable most sensitive to noise, increasing during exposure with no evidence of habituation. Systolic blood pressure was unaffected by exposure to noise. Heart rate decreased the most during exposure to irregular noise. Most interestingly, HR oscillations centering around the 0.1 Hz frequency were decreased following exposure to noise. HR oscillations at this frequency have been attributed to the spontaneous oscillations of the blood pressure control system, suggesting that noise influences this control system. Finally, many of the hemodynamic effects were greatest late in the protocol, a finding which led the authors to surmise that previous laboratory studies which found no significant effect of noise exposure were too short to detect effects. The fact that responses to noise did not show a habituation effect lends some credibility to the putative chronic effects of noise exposure.

Apart from the possible direct effects of noise on cardiovascular disease, one recent study examined the indirect effect via changes in smoking behavior, a known risk factor for heart disease. Cherek (1985) examined smoking while subjects performed an operant task during exposure to different levels of industrial noise (70, 80, and 90 dB). As noise levels increased, so did smoking.

Thus, the most consistent findings emerging from recent human laboratory research on the cardiovascular effects of noise relate to blood pressure regulation. While negative studies exist, it appears that at least in the laboratory, short-term exposure to loud noise reliably produces increases in diastolic blood pressure. Moreover, several recent studies have provided evidence that although the mechanisms are not fully understood, these effects are mediated through alterations of the central blood pressure control system.

Studies of Animal Subjects. Animal studies represent something of a compromise between the limitations of laboratory and field studies in that they permit controlled exposure to noise in studying the long-term development of cardiovascular disease. As early as 1945, exposure to random noise was observed to produce elevations in blood pressure in laboratory rats (Medoff and Bongiovanni 1945). Borg (1978) demonstrated a dose-response effect of noise in tail artery vasoconstriction in rats. Nevertheless, other studies (Borg and Møller 1978; Pillsbury 1986) using normotensive and spontaneously hypertensive rats have demonstrated that exposure to white noise pulses of random length did not produce elevations in blood pressure at lengthy follow-up assessments.

However, as DeJoy (1984) has indicated, the choice of the rat as a model for the study of the cardiovascular effects of noise may not be appropriate. Not only are there differences in the structure and function of the auditory system between the human and the rat but the latter is also susceptible to audiogenic seizures.

Peterson and colleagues (1981, 1984) have conducted a series of studies using a primate model to assess the effects of noise on the cardiovascular system. Peterson et al. (1981) demonstrated that in rhesus monkeys, nine months of daily exposure to levels of noise comparable to that produced in some industries led to sustained increases in blood pressure. Moreover, this increase was not reversed in the month following exposure. In addition, these increases in blood pressure were not accompanied by hearing loss.

More recently, Peterson et al. (1984) addressed a problem in the design of the 1981 study: that the post-exposure period was not sufficient to assess whether the elevations in blood pressure would return to baseline values. In this study, two macaque monkeys were allowed nine days in which their baseline pressures were ascertained; there followed 97 days of exposure to high levels of noise, and a 118-day post-exposure period. The noise stimulus, as in the previous study, was a recorded sequence approximating the noise patterns of a noisy work environment. Animals were exposed to four hours per day of noise until their blood pressure reached a plateau. At this point, exposure to the noise stimulus increased to eight hours per day.

Heart rate increases during the exposure period were nonsignificant but across the entire exposure mean arterial pressure (MAP) rose from 85 ± 2.4 to 94 ± 2.5 mm Hg, a significant increase. Moreover, exposure to four hours of noise per day led to a MAP increase of 8.2% while exposure to eight hours per day produced an increase of 16.5%, suggesting a dose-response effect. During the first 30 days of the post-exposure period, MAP decreased from the peak of the exposure period but remained elevated compared to baseline values.

Other primate studies have reported less robust findings regarding the effects of noise on cardiovascular outcomes. Kraft-Schreyer and Angelakos (1979), studying African green monkeys, were unable to detect consistent blood pressure responses to noise exposure lasting up to four months. Turkkan, Hienz, and Harris (1984) reported that long-term exposure to differing levels of noise had an acute dose-response effect on blood pressure, both systolic and diastolic, as well as on heart rate. However, responses appeared to habituate over time. Blood levels of epinephrine and norepinephrine also decreased after exposure to noise compared to baseline values, suggesting that noise led to a decrease in activity of the sympathetic system.

Kirby et al. (1984) reported that a single exposure to 30 minutes of 95 dB broadband noise led to significant increases in mean arterial pressure in monkeys with hypertensive parents but not in the offspring of normotensive parents. These elevations were significant in spite of significant baseline differences in blood pressure. The authors speculate that these effects may be due to enhanced sympathetic and vascular reactivity in the offspring of hypertensive animals.

In summary, recent animal research on the cardiovascular effects of noise exposure has focused most consistently on blood pressure as the outcome variable. Apart from this, however, there is little consistency. Studies using a rat model sometimes demonstrate a blood pressure-increasing effect of noise and sometimes do not. However, the choice of the rat as a model may not be appropriate, as DeJoy (1984) has suggested.

Research in primates is slightly more consistent. Most studies show an effect of noise on blood pressure, usually measured as mean arterial pressure, and there is evidence to support a dose-response effect. Questions remain, however, regarding the degree to which this effect habituates with time. The work of Peterson and colleagues suggests that habituation does not occur (1984). Other studies suggest the opposite. Differences in methodology as well as in selection of primate subjects may account for some of this discrepancy.

Field Studies

Early epidemiological studies often reported detrimental effects of noise on the cardiovascular system. Thompson (1981), after reviewing over 70 of these field studies, concluded that the data were not strong enough to permit unequivocal assertions about the effects of long-term exposure to noise on the cardiovascular system. He also reported that the vast majority of these field studies examined the relationships between exposure to noise and the development of hypertension. In these studies, most of which were cross-sectional, there appeared to be some consistency in finding an association between noise exposure and elevated blood pressure but the differences between exposed and unexposed groups was small.

Traffic Noise. Since Thompson's review was published, several other field studies have appeared. The Caerphilly Study (Babisch et al. 1988) assessed the relationship between traffic noise and cardiovascular risk using a cross-sectional design. A noise map of the area under study was constructed, resulting in assessments of traffic noise emission from 6 A.M. to 10 P.M. measured 10 m from the center of the street, as well as noise immission measured at the facades of the homes in the area. A wide variety of indicators of cardiovascular risk was collected by questionnaire and physical examination of 2,512 men in the area, representing 89% of the eligible men between the ages of 45 and 59 years. By and large, the results were negative. There was no association of ischemic heart disease prevalence and traffic noise, nor was there evidence of associations between noise exposure and diastolic blood pressure, heart rate, various measures of clotting, and levels of very low-density lipoprotein (VLDL) cholesterol, low-density lipoprotein (LDL) cholesterol, and triglycerides. The authors report associations between noise and systolic blood pressure, plasma viscosity, levels of total cholesterol, high-density lipoprotein (HDL) cholesterol, testosterone, and cortisol, among other measures, but not all of these were in the direction associated with increased risk. Moreover, some of these associations were curvilinear, with lowest risk associated with intermediate exposure to noise. Finally, as the authors themselves point out, since the area under study was suburban in character, high levels of traffic noise were relatively low (70 dBA) compared to those in field studies of other types of noise exposure. Overall, the findings of this part of the Caerphilly study are equivocal.

Neus et al. (1983) also have reported on the health impact of exposure to traffic noise. Unlike the Caerphilly study, theirs found increased incidence of hypertension in areas of high noise compared to low noise. Moreover, in their high noise areas, noise levels reached only 66 dBA compared to 51 dBA in the quiet areas. Both of these values are exceeded by the high-noise levels in the Caerphilly study.

Industrial Noise. In another recent cross-sectional field study (van Dijk, Verbeek, and de Fries 1987), 278 male employees in the shipbuilding (high noise, average level = 98 dBA) and machine shop (low noise, average = 85.5 dBA) departments of the Dutch Royal Navy Shipyard were studied. Data recording blood pressure, height, weight, and audiometric assessment were collected during a physical examination. Although total years of exposure and age of the two groups were similar, no difference in SBP or DBP was found, either before or after correction for age. However, in the same work, the authors also report on a study of 428 male employees exposed to noise levels over 89 dBA in the production departments of six factories. Physiological data were collected during a brief physical examination. A final analysis was carried out on 297 of these employees. After correction for age, those employees exposed for at least 20 years had SBP and DBP values higher than employees exposed for fewer than 10 years. Moreover, in the high exposure group, prevalence of hypertension was greater (28%) than in the low exposure group (9%). However, no statistical tests of differences are provided by the authors, nor is there a description of factors that led to a substantial subject drop-out rate. Finally, no information is available as to either body-mass index or the number of employees on anti-hypertensive medication.

Wu, Ko, and Chang (1987) also conducted a cross-sectional study of exposure to noise in a shipyard. After screening 3,746 employees, 158 workers from a high noise work environment (> 85 dBA) were matched with 158 workers from a low noise environment (< 80 dBA) for age, months of employment, and body-mass index. Data were collected during physical examinations. SBP and DBP of the exposed group were significantly higher ($p < .01$) than in the low noise group. Prevalence of hypertension was also higher in the exposed group. As expected, the groups did not differ with respect to the matching variables.

Belli et al. (1984) report no relationship between exposure to high frequency noise and cardiovascular outcomes. They studied prevalence of hypertension among 940 workers: 579 employed in a textile plant and 361 employed in municipal offices. These two populations did not differ with respect to age, but there was no evaluation of other potentially confounding variables such as weight or length of time of exposure. Of the textile workers, 490 were exposed to average noise levels greater than 85 dBA. The remaining textile workers as well as the municipal employees represented the nonexposed group ($N = 450$). Blood pressure was measured during a medical examination. The prevalence of hypertension, defined by World Health Organization (WHO) criteria, was 17% in the exposed group compared to 14% in the nonexposed group, a nonsignificant difference.

Aircraft Noise. Kent, von Gierke, and Tolan (1986) retrospectively studied the medical records of 2,250 Air Force aircrew members referred to the School of Aerospace Medicine for evaluation of conditions which had the potential to disqualify them from flying. High-frequency hearing loss was used as an estimate of exposure to occupational noise. Cardiovascular measures included SBP, DBP, essential hypertension, secondary hypertension, and coronary artery disease (CAD). After controlling for age, no significant effect of hearing loss was found for any of these measures. Relative risk of CAD also was not influenced by hearing loss. The authors indicate that this study population was not representative of the civilian population exposed to occupational noise and that, in fact, they were not even representative of other aircrew members since they had some condition that required referral for medical evaluation. Moreover, the measurement of exposure in this study is questionable since cardiovascular effects such as blood pressure elevations can be produced by noise exposure even in the absence of actual impairment of hearing (Peterson et al. 1981).

In one of the more carefully controlled studies in the literature, Cohen et al. (1981) have reported on the consequences of exposure to aircraft noise in children. Subjects were children attending the noisiest elementary schools in the vicinity of the Los Angeles airport, with over 300 overflights per day and with noise levels reaching 95 dBA. Control subjects were students from schools matched for grade level, ethnic and racial distribution, and other potentially confounding variables. While the matching was imperfect, in that differences between the two samples were found in racial distribution and mobility, regression analyses statistically controlled for these differences. Children from the noisy schools had higher systolic and diastolic blood pressure than control subjects. A marginally significant interaction between years of exposure to the noisy classrooms and systolic blood pressure suggested a tendency toward habituation. No such tendency was found for diastolic blood pressure.

However, Watkins et al. (1981, cited by Cohen et al. 1986) found no effect of aircraft noise exposure on use of prescription medications or general medical services, as might be expected if there were a relationship between noise exposure and hypertension. Von Eiff, Friedrich, and Neus (1982), studying communities in the vicinity of the Munich airport, found that at low to moderate levels of aircraft noise, there was an increase in the incidence of hypertension. High levels, however, were associated with higher levels of blood pressure.

To summarize, examination of numerous field studies of the effect of noise on the cardiovascular system published since Thompson's review reveal inconsistent findings. Like most of the laboratory studies, these investigations tend to focus on blood pressure responses to noise, although some assess other cardiovascular variables. Some of the more recent studies are more sound in their design than their predecessors in that they address potentially confounding variables and compare low noise to high noise conditions. In general, although there are inconsistencies in the findings of these recent studies, they generally support the assertion that exposure to noise is associated with higher levels of blood pressure.

Conclusions

Overall, while there are inconsistencies, careful reading of the recent field and laboratory studies examining the cardiovascular effects of exposure to noise suggests that (1) exposure to noise is associated with elevations in blood pressure, primarily diastolic blood pressure, and (2) that in general, this effect appears not to habituate over time. These findings regarding diastolic blood pressure responses are strengthened by recent evidence suggesting noise-induced vasoconstrictive responses and effects on the central blood pressure control system. Apart from blood pressure, there is little evidence for effects of noise on other cardiovascular health outcomes. Regrettably, even these assertions about hypertension cannot be stated unequivocally since significant negative findings can be found in the literature.

In addition to the inconsistencies in the recent literature cited above, certain methodological considerations arise with regard to potential mediating factors not yet fully studied. For example, as Kryter (1985) and Neus, Rüddel, and Schulte (1983), among others, have suggested, the subjective evaluation of noise may be an important variable mediating the relationship between health outcomes and exposure to noise. The degree to which this is true is unknown since relatively few studies have examined this potentially important variable.

Secondly, noise itself is a complex stimulus which, in addition to intensity, varies in frequency, range, and duration (intermittent versus continuous). While it is possible to characterize precisely the qualities of a noise stimulus in the laboratory or even in an industrial or community setting, it becomes an enormous task to vary these characteristics systematically in order to determine which are the most important. Finally, the degree of perceived control and predictability of noise may be an important mediating factor in determining cardiovascular health outcomes (Cohen et al. 1986).

The conclusion may be drawn that our knowledge of the cardiovascular consequences of exposure to noise is still limited. While the data seem to point to a blood pressure-elevating effect of noise, there are inconsistencies in the literature that remain unresolved. Clarification of the role of subjective responses to noise stimuli, the characteristics of the stimuli themselves, and the degree of perceived control and predictability of the noise may help in resolving some of these inconsistencies.

REFERENCES

Abbey-Wickrama, I., a'Brook, M. F., Gattoni, F. W. G., and Herridge, C. F. 1969. Mental hospital admissions and aircraft noise. *Lancet* 2:1275-1277.

Andrén, L., Hansson, L. Björkman, M., Jonsson, A. 1980. Noise as a contributing factor in the development of elevated arterial pressure. *Acta Medica Scandinavica* 207:493-498.

Andrén, L., Lindstedt, G., Björkman, M., Borg, K. O., and Hansson, L. 1982a. Effect of noise on blood pressure and "stress" hormones. *Clinical Science* 63:3715-3745.

Andrén, L., Piros, S., Hansson, L., Herlitz, H., and Jonsson, O. 1982b. Different hemodynamic reaction patterns during noise exposure in normotensives with and without heredity for hypertension. *Clinical Science* 62:137-141.

Andrén, L., Hansson, L., Eggertsen, R., Hedner, T., and Karlberg, B. E. 1983. Circulatory effects of noise. *Acta Medica Scandinavica* 213:31-35.

Babisch, W., Ising, H., Gallacher, J. E. J., and Elwood, P. C. 1988. Traffic noise and cardiovascular risk. The Caerphilly study, first phase. Outdoor noise levels and risk factors. *Archives of Environmental Health* 43:407-414.

Belli, S., Sani, L., Scarficcia, G., and Sorrentino, R. 1984. Arterial hypertension and noise: A cross-sectional study. *American Journal of Industrial Medicine* 6:59-65.

Borg, E. 1978. Peripheral vasoconstriction in the rat in response to sound. I. Dependence on stimulus duration. *Acta Otolaryngologica* 85:153-157.

Borg, E., and Møller, A. R. 1978. Noise and blood pressure: Effects of life-long exposure in the rat. *Acta Physiologica Scandinavica* 103:340-342.

CARTER, N. L., and BEH, H. C. 1989. The effect of intermittent noise on cardiovascular functioning during vigilance task performance. *Psychophysiology* 26:548-559.

CHEREK, D. R. 1985. Effects of acute exposure to increased levels of background industrial noise on cigarette smoking behavior. *International Archives of Occupational and Environmental Health* 56:23-30.

COHEN, S., and WEINSTEIN, N. 1981. Non-auditory effects of noise on behavior and health. *The Journal of Social Issues* 37:36-70.

COHEN, S., KRANTZ, D. S., EVANS, G. W., and STOKOLS, D. 1981. Cardiovascular and behavioral effects of community noise. *American Scientist* 69:528-535.

COHEN, S., EVANS, G. W., STOKOLS, D., and KRANTZ, D. S. 1986. *Behavior, Health, and Environmental Stress.* New York: Plenum.

COLLETTI, V., and FIORINO, F. G. 1987. Myocardial activity during noise exposure. *Acta Otolaryngologica* 104:217-224.

DEJOY, D. M. 1984. The nonauditory effects of noise: Review and perspectives for research. *Journal of Auditory Research* 24:123-150.

EGGERTSEN, R., ANDRÉN, L., and HANSSON, L. 1984. Haemodynamic effects of loud noise in hypertensive patients treated with combined beta-adrenoceptor blockade and precapillary vasodilatation. *European Heart Journal* 5:556-560.

EGGERTSEN, R., SVENSSON, A., MAGNUSSON, M., and ANDRÉN, L. 1987. Hemodynamic effects of loud noise before and after central sympathetic nervous stimulation. *Acta Medica Scandinavica* 221:159-164.

KENT, S. J., VON GIERKE, H. E., and TOLAN, G. D. 1986. Analysis of the potential association between noise-induced hearing loss and cardiovascular disease in USAF aircrew members. *Aviation, Space, and Environmental Medicine* 57:348-361.

KIRBY, D. A., HERD, J. A., HARTLEY, L. H., TELLER, D. D., and RODGER, R. F. 1984. Enhanced blood pressure responses to loud noise in offspring of monkeys with high blood pressure. *Physiology and Behavior* 32:779-783.

KRAFT-SCHREYER, N., and ANGELAKOS, E. T. 1979. Effects of sound stress on norepinephrine responsiveness and blood pressure. *Federation Proceedings* 38:883 (Abstract).

KRYTER, K. 1985. *The Effects of Noise on Man.* 2d. ed. Orlando: Academic Press.

LAIRD, D. A. 1932. Experiments on the influence of noise upon digestion and the counteracting effects of various food agencies. *Medical Journal and Record* 135:461-464.

LINDEN, W. 1987. Effect of noise distraction during mental arithmetic on phasic cardiovascular activity. *Psychophysiology* 24:328-333.

MEDOFF, H. S., and BONGIOVANNI, A. M. 1945. Blood pressure in rats subjected to audiogenic stimulation. *American Journal of Physiology,* 143:300-305.

NEUS, H., RÜDDEL, H., SCHULTE, W., and VON EIFF, A. W. 1983. The long-term effect of noise on blood pressure. *Journal of Hypertension* 1(Supplement 2): 251-253.

NEUS, H., RÜDDEL, H., and SCHULTE, W. 1983. Traffic noise and hypertension: An epidemiological study on the role of subjective reactions. *International Archives of Occupational and Environmental Health* 51:223-229.

PETERSON, E. A., AUGENSTEIN, J. S., TANIS, D. C., and AUGENSTEIN, D. G. 1981. Noise raises blood pressure without impairing auditory sensitivity. *Science,* 211:1450-1452.

PETERSON, E. A., AUGENSTEIN, J. S., HAZELTON, C. L., HETRICK, D., LEVENE, R. M., and TANIS, D. C. 1984. Some cardiovascular effects of noise. *Journal of Auditory Research* 24:35-62.

PILLSBURY, H. C. 1986. Hypertension, hyperlipoproteinemia, chronic noise exposure: Is there synergism in cochlear pathology? *Laryngoscope* 96:1112-1138.

RAY, R. L., BRADY, J. V., EMURIAN, H. H. 1984. Cardiovascular effects of noise during complex task performance. *International Journal of Psychophysiology* 1:335-340.

SMITH, E. L., and LAIRD, D. A. 1930. The loudness of auditory stimuli which affect stomach contractions in healthy human beings. *Journal of the Acoustical Society of America* 15(1):94-98.

THOMPSON, S. J. 1981. Epidemiology feasibility study: Effects of noise on the cardiovascular system. United States Environmental Protection Agency. Washington, DC.

TURKKAN, J. S., HIENZ, R. D., and HARRIS, A. H. 1984. Novel long-term cardiovascular effects of industrial noise. *Physiology and Behavior* 33:21-26.

VAN DIJK, F. J. H., VERBEEK, J. H. A. M., and DE FRIES, F. F. 1987. Non-auditory effects of noise in industry. *International Archives of Occupational and Environmental Health* 59:55-62.

VON EIFF, A. W., FRIEDRICH, G., and NEUS, H. 1982. *Traffic Noise, a Factor in the Pathogenesis of Essential Hypertension.* Contributions to Nephrology, vol. 30, 82-86, eds. G. M. Berlyne, S. Giovanetti, and S. Thomas. Basel: S. Karger.

WU, T-N., KO, Y-C., and CHANG, P-Y. 1987. Study of noise exposure and high blood pressure in shipyard workers. *American Journal of Industrial Medicine* 12:431-438.

Neuroendocrine, Immunologic, and Gastrointestinal Effects of Noise

LAWRENCE W. RAYMOND

That excessive noise can cause deafness is beyond question. Noise-induced hearing loss is a major social problem, and may be the most common occupational disease (van Dijk, Ettema, and Zielhuis 1986). Other, non-auditory effects of noise, however, have received little attention (or their study has left major questions unanswered). The aim of this review is to examine the evidence for such effects, across the broad range of clinical and experimental observations in human subjects, as of 1990. Where appropriate, we shall target those areas in which further investigation is needed (van Dijk, Ettema, and Zielhuis 1987).

Summary of Cardiovascular Findings

There is a substantial but confusing literature in this area, most of which addresses the putative role of noise in the common problem of hypertensive disease. A myriad of study designs have been brought to bear on this important question, including observations of acute and chronic conditions in several animal species and of acute responses of normal human subjects and patients as well, and epidemiological perspectives of subjects exposed to either aircraft, occupational, or road-traffic noise. The highly variegated nature of noise *per se*—intensity, duration, acoustic frequency, intermittency, and other features—has greatly complicated the deliberations.

Furthermore, since noise rarely occurs as an isolated stressor outside the laboratory, many investigators have chosen to couple it with other physical or mental perturbations in examining blood pressure and related responses to the aggregate stimulus.

Even the object of study—blood pressure—is not an altogether stable concept. Systemic hypertension admits of variations in definition across the international body of researchers as to its degree, duration, and physiopathologic impact, and the means of measurement of blood pressure undergo periodic reexamination (Frolich 1988).

Given the wide-ranging nature of the data base, it is not surprising that only limited generalizations can be made, of which the following may be representative:

1. Abnormal blood pressure levels can be induced by high noise levels in some animal species, and in humans under some laboratory conditions. The effects in animals have been long-lasting in some studies.
2. The relevance of these findings to human subjects is unclear. For example, it is not known whether controlling noise to levels which do not usually cause hearing loss (e.g. < 85 dBA, 40-hour/week exposures) leaves any residual risk of causing hypertension, except perhaps in unusually sensitive subjects such as those identified by Petiot et al. (1988).
3. If such subjects could be reliably identified, lower levels of noise might be tolerated without a pressor response, but the intensity-duration relationship has not been studied.

In discussing some of the studies of chronic exposure to noise, Rosenman (1990) noted an association of such exposure with an increased incidence of hypertension, whether or not hearing loss was also demonstrated. In the study by Fouriaud and colleagues (1984), such loss was not even considered, and the criterion for noise exposure was apparently established only with reference to a one-time measurement by a physician (≥ 85 dBA). Classification difficulties were also present in the study of blue-collar factory workers by Talbott and her associates (1985). They compared workers from a noisy plant (≥ 89 dBA) with those of a less noisy plant (< 81 dBA), and found no difference in their mean systolic or diastolic blood pressures. However, 40% of the workers in the noisy plant said they almost always wore hearing protection, and at least 17% of workers in the less noisy plant had formerly worked in noisy press areas (where no noise surveys were permitted). Nevertheless, Talbott et al. did find a strong relationship between severe high-frequency hearing loss (e.g., losses ≥ 65 dBA) and hypertension in persons over age 56, regardless of where they worked. In contrast to the view of Rosenman (1990), the conclusion reached, after reviewing other studies, by Horvath and Bedi (1990) in the same publication, was that there is little evidence that, except for hearing loss, loud noise produces any functional alteration other than a transient stress reaction. This view is shared by Abel (1990).

While the influence of noise on blood pressure remains to be further clarified, other possible effects of noise on cardiovascular function and disease may be equally important. Effects such as changes in levels of catecholamines and cholesterol, in platelet function, and in other risk factors are discussed below.

As is often the case in scientific inquiry, one is left with the impression that we do not yet know enough to draw firm general conclusions. Do we need more research? How best can we design further studies so as to take advantage of what is already known? Surely a multidisciplinary approach to future study designs offers the brightest hope for continuing progress.

Neuroendocrine and Other Metabolic Effects

Research findings in this area have recently been summarized. Van Dijk, Ettema, and Zielhuis (1986) offer the general hypothesis that the reticular arousal system is the central focus for noise-induced effects, so that activation produced both in the higher cortical centers and in the hypothalamus in turn activates not only the adrenal medulla by autonomic pathways, but also the adrenal cortex by increased hypophyseal secretion of adrenocorticotropin (ACTH). Thus, increases in catecholamines and cortisol concentrations in peripheral blood are elicited.

Arguelles and his colleagues (1970) have reviewed their earlier work (Arguelles et al. 1962) in which increases in epinephrine (E) and norepinephrine (NE) were consistently ($p < .001$) observed upon exposure to noise in normal subjects as well as in patients who had recovered from heart attacks. Plasma cortisol and cholesterol levels, already higher in the latter group, were unchanged by three-hour exposures to 90 dBA noise. The later study (Arguelles et al. 1970) reports very similar findings with patients who had labile hypertension, in whom noise caused a further rise in blood pressure (from 160/100 to 170/110, $p < .01$), in urinary E ($p < .05$), and in NE ($p < .001$), while high baseline levels of plasma cortisol and cholesterol were not further raised. This lack of further elevation is not inconsistent with the earlier finding of Arguelles and his coworkers (1962) that ACTH rises after exposure to noise in normal human subjects. In the above patients (Arguelles et al. 1970) ACTH secretion may have already been maximally stimulated—it was not actually measured in this more recent study. The speed of the ACTH response is impressive. A recent German study of young volunteers exposed to 105 dBA for 10 seconds in an aircraft simulator found a rise in ACTH in all 25 subjects which reached "pathological levels" in seven of them (Marth et al. 1988). No sham exposures appear to have been included, to test for the role of anticipation in the observed response.

The rapidity of the pituitary-adrenal response has been elegantly demonstrated in awake dogs chronically equipped with adrenal vein and femoral artery catheters (Engeland, Miller, and Gann, 1990). Using a three-minute signal from a laboratory timer (75 dB, 0.25-8 kHz band width), they showed brisk increases in ACTH concentrations immediately followed by a rise in cortisol secretion. Secretion of E and NE also rose promptly to levels three times their baseline levels, falling back to baseline within three minutes after noise ceased. ACTH concentrations and cortisol secretion returned more slowly to their pre-noise values (10 to 15 minutes).

More chronic but intermittent noise produced more sustained effects on levels of blood cortisol and cholesterol in the classic study of Cantrell (1974). He exposed each of 20 military men to 80-, 85- and 90-dB tonal pulses in the 3-4 kHz range for a period of 10 days in a row. Cortisol levels rose from 12.2 up to 21 mcg/deciliter, higher rises occurring at the higher noise levels, and remained elevated (17.5 mcg/dL, $p < .001$) until the experiment ended, 15 days after noise was discontinued. Cholesterol values showed a similar (but less striking) rise, from 175 mg/dL before exposure to 192-215 mg/dL upon exposure, and fell to 193 mg/dL 15 days after exposure. Cantrell's data indicate that these metabolic changes do not show evidence of adaptation, and furthermore, it seems clear that they are unlikely to be attributable to confinement (1974).

Other observations supporting the view that catecholamines play a key role in noise-induced responses include several German studies in which the excretion of both E and NE (with E predominating) was increased when noise was combined with a mental task (Klotzbücher 1976; Klotzbücher and Fichtel 1979; Ising 1980). However, Cesana (1982a) found noise-related increases only in urinary NE in his study of experienced Italian pressmen, among whom E excretion seemed to be more related to subjective perception of chronic features of rotating shift work. Urinary dopamine (DA) excretion bore little relation to shift work or noise (Cesana 1982b).

Another industrial study involved normal men in a glass factory, of whom 60 worked amid noise over 90 dBA and were compared with 52 coworkers exposed to ambient noise below 78 dBA (Cavatorta et al. 1987). Post-work levels of plasma E and NE, and their urinary metabolite, vanillylmandelic acid, showed a significant rise ($p < .01$), both in comparison to their pre-work baselines, and in relation to the post-work levels of the low-noise group. Plasma concentrations of DA were unchanged in both groups, consistent with the results of Cesana (1982b) with respect to urinary levels. Blood cortisol and urinary homovanillic acid levels were also unchanged in the noise-exposed group.

Catecholamine responses to brief periods of traffic noise have been observed by Catapano and his colleagues (1984), who found that plasma NE rose more significantly than E or DA as early as five minutes into the exposure, which raised diastolic blood pressure without affecting systolic blood pressure or heart rate.

Not all observations concur. Earlier studies by van Dijk, Souman, and de Vries (1983) and by Osguthorpe and Mills (1982) found no change in catecholamine excretion in normal volunteers exposed to 98 or 84-90 dBA, respectively. An increase in plasma cortisol was seen in the latter group, persisting 16 hours after the end of noise exposure in the 90-dBA subjects. Cortisol was not measured by van Dijk, Souman, and de Vries (1983). Completely negative results were reported by Andrén, Hansson, and Björkman (1981), who exposed healthy men to 95 dBA noise for 20 minutes. They found no changes in plasma concentrations of E, NE or cortisol, despite a 12% rise in diastolic and mean blood pressure ($p < .001$). There was no change in systolic blood pressure or heart rate. Similarly, negative results for E, NE, and DA were noted by Brandenberger and his associates (1980), whose subjects carried out a short-term memory task under either noisy (99 dBA) or quiet (45 dBA) conditions. No further rise in E, NE or DA was elicited by noise, beyond that caused by the task under quiet conditions. However, the task-related rise in plasma cortisol was greater under noisy than under quiet circumstances—noise may have been responsible for one quarter of the overall rise in cortisol. Surprisingly, the rise in cortisol was not attributable to changes in blood levels of ACTH, perhaps because of problems with the assay, which used a commercial ACTH antiserum. In a more recent study from the same laboratory (Wittersheim, Brandenberger, and Follenius 1985), blood cortisol changes in relation to similar experimental influences were examined, but data relating ACTH changes with the cortisol increases they observed were not reported.

The few studies of noise effects on other pituitary hormones have also produced inconsistent results. The normal rise in serum growth hormone (GH) associated with the onset of sleep was steeper in three of six subjects after eight-hour exposures to 83 dBA of pink noise (Fruhstorfer et al. 1985). Similar results for serum prolactin (PRL) were found in the same three subjects. The earlier work of Brandenberger and his colleagues (1980) showed no GH response. The lack of a noise effect on GH was also reported by Andrén, Hansson, and Björkman (1981), who also found no effect on PRL.

It is unclear why three of the six subjects of Fruhstorfer et al. (1985) reacted to noise with increased GH and PRL responses. Only one of these 6 subjects showed an increase in blood ACTH level after exposure to noise. However, the six generally showed a noise-induced reduction in nocturnal levels of tryptophan, serotonin, and their metabolite, 5-hydroxyindoleacetic acid (5-HIAA), an effect which the authors interpreted as due to an intensification of recovery processes by way of compensating for prior oversecretion during exposure to noise on the previous day. They cite animal experiments that provide some support for this speculation, but direct measurements of tryptophan, serotonin, and 5-HIAA in relation to noise exposure in humans have not yet been reported.

One of the earliest physiological variables to be studied with regard to noise is the blood glucose level. As cited by Loeb (1986), studies in human subjects and subhuman primates have found increases in blood sugar, but the mechanisms responsible for the increases have not been adequately identified. No comprehensive study of the effect of noise on glucose-insulin coupling or its relation to counter-regulatory hormones has yet been published. One report on brief exposure to traffic noise found no change in blood glucose concentrations, nor in cyclic adenosine monophosphate or uric acid (Catapano et al. 1984).

Far too little attention has been paid to possible effects of noise on blood lipoproteins, despite a potential link both to noise-induced hearing loss, and to the widely-recognized condition of atherosclerosic cardiovascular disease. Resnekov (1981) and Axelsson and Lindgren (1980) summarized information available around a decade ago (the former's work was part of an American Heart Association task force on "Environment and the Cardiovascular System"). The most consistent finding, and probably the most important, is the increase in total serum cholesterol observed when human subjects or smaller mammals are exposed to noise at 80-90 dBA, or higher. It would seem to be an important next step to investigate which components—especially the undesirable low-density lipoproteins (LDL) and the protective high-density lipoproteins (HDL)—are altered by noise, and to discover whether the changes are sustained with continuing noise exposure. An interesting beginning has been made in this regard by Babisch and his coworkers (1988) in their study of possible cardiovascular risks associated with traffic noise. Even over the range of 51-70 dBA, a positive association of noise with total cholesterol has been observed, along with systolic blood pressure, blood estradiol and antithrombin III concentrations, and plasma viscosity. A negative correlation of noise with plasma cortisol level and with blood platelet count was observed over the modest noise range thus far studied. In future work, occupational noise exposures are to be considered, and noise effects on blood lipoprotein subfractions may emerge as being of greater importance.

Even ultra-short noise exposure has been shown to increase serum cholesterol levels in normal volunteers by Marth and his associates (1988), who used a recorded overflight in which the sound pressure level rose from 25 dBA to a three-second peak of 105 dBA, and fell back to 25 dBA, all in less than 10 seconds. Total cholesterol increased in 85% of subjects ($p < .01$) from a mean of 146 mg to 156 mg per deciliter. A smaller but consistent fall in triglycerides also occurred, while changes in blood glucose, free fatty acid and hepatic transaminase concentrations were minimal.

Such a rise in cholesterol (or in lipoprotein subfraction) could be the cause of noise-induced platelet aggregation, which has been observed in rabbits and rats, according to Russian and German investigators cited by Resnekov (1981). The potential public health implications of these observations are substantial, calling for further experiments in human subjects who exhibit both normal and abnormal lipoprotein patterns. At the same time, it should be borne in mind that one investigation found cholesterol and triglyceride levels to be unaffected by five- to ten-minute traffic-noise exposures which were sufficient to increase catecholamine concentrations and diastolic blood pressure (Catapano et al. 1984).

Data on noise effects upon thyroid function are quite sparse. Jonderko and his colleagues (1982) studied Polish construction workers who performed hard physical labor in quiet circumstances and in conditions involving whole-body vibration and exposure to noise (90-108 dBA). They found that noise blunted the thyroid-stimulating effect of hard work, an effect which was undampened by vibration. Work appeared to cause increases in circulating levels of thyroid hormones through increases in thyroid-stimulating hormone (TSH). No measurements of TSH-releasing hormone in relation to TSH and thyroid hormones in noise-exposed subjects have been reported.

Gastrointestinal Effects

Information on possible noise-induced changes in gastrointestinal function has been characterized as inconclusive (Westman and Walters 1981) in spite of some earlier suggestions that such effects are important consequences of occupational exposures (Guilian 1974; Cohen, Glass, and Phillips 1979).

Influences on gastrointestinal motility have been shown in two ways. Increased gastric emptying was produced by noise in subjects studied by Kaus and Fell (1984). More recently, Young and his colleagues (1987) combined 100 dBA of white noise with either a cold-pressor or cognitive-task challenge. In both instances, peristaltic esophageal contractions increased in amplitude, as did self-reported anxiety. Esophageal contraction velocity increased only in the experiments in which noise was combined with a cognitive exercise. The effect of noise alone on esophageal motility was not evaluated.

In contrast, no change was observed in gastric acid secretion or mucosal blood flow in subjects exposed to 90 dBA of broad-band noise applied by headphones (Sonnenberg et al. 1984). The stimulus was potent enough to elicit small but consistent increases in both systolic and diastolic blood pressure in a pilot study, and a significant rise in systolic blood pressure (18 mm Hg, $p < .05$) in the gastric-response study. Unpublished observations from this same laboratory have suggested that 90 dBA noise may alter small-bowel transit time, stool frequency, and stool volume, under laboratory conditions.

In a study of aircraft noise in the Netherlands, Knipschild (1977) found evidence of an excess of cardiovascular and mental illness, but gastrointestinal symptoms were minimally affected. Use of prescription antacids, however, did appear to increase with increases in noisy flight operations in a noise-exposed village, but their use remained stable in a similar (control) village. This finding may involve an understatement of the possible association, since the use of non-prescription antacids or dietary modifications would not have been captured by the study's design (Knipschild and Oudshoorn 1977). The increased use of prescription antacids was accompanied by increases in the use of hypnotics, sedatives, antihypertensives, and other cardiovascular agents in the noisy, but not in the control community. That the increase in antacid and in antihypertensive agents continued after night flights were curtailed (van Dijk, Ettema, and Zielhuis 1986) raises the possibility of sustained disturbances of gastrointestinal and cardiovascular function, perhaps linked in some way. As Loeb (1986) stated, "The case for noise, as a stressor, in increasing the incidence of certain diseases and syndromes, especially cardiovascular and gastric disorders, is compelling but not conclusive." Indeed, the case for gastrointestinal dysfunction has not been made, in part because of confounding factors, but also perhaps on account of inadvertence on the part of occupational physicians and other clinicians. With regard to noise-related effects, the gastrointestinal tract appears to be a large terra incognita at this time.

Although noise was not included in a list of perturbations expected in spaceflight reported by Pfeiffer (1968), environmental conditions that were named, along with their effects on the gastrointestinal tract, including such factors as acceleration, vibration, heat, weightlessness, decompression, hypoxia, oxygen toxicity, fatigue, and psychological stress. It is not illogical to suggest that noise might adversely affect gastrointestinal function as readily as some of these factors. Future researchers on digestive processes in earthly environments might find Pfeiffer's report to be of heuristic value.

Hematologic and Immunologic Effects

Amid an explosive expansion of the field of immunology and the advent of psychoimmunology, one finds very little information on the physiological effects of noise on the immune system. Lisiewicz and Moszczyzinski (1986) reviewed the published reports, many of them in the Polish literature (and described to me in Suczewski 1989), incorporating additional environmental perturbations such as vibration, thermal stress, and the presence of toxic substances. With regard to the formation of red blood corpuscles (erythropoiesis), they found the effect to be small and of limited clinical relevance. Effects of noise on neutrophil numbers and on enzymatic content or function appear to be a matter of some inconsistency and controversy, as is true of what few studies exist on the formation of lymphocytes or lymphoid tissue (lymphopoiesis). One finding that seems consistent among several groups is that noise elicits a rise in circulating lymphocytes so young that they do not yet possess the normal array of enzymatic contents. Serum immunoglobulin (Ig) concentrations have been found to drop in workers exposed to noise, and IgG, IgA and IgM levels were lower, the more intense the noise. The addition of vibration appears to oppose this effect, and no change in the above immunoglobulins was found in deaf workers exposed to noise. Further work seems needed to enable a consistent picture to emerge, as well as to address possible effects on IgE levels, which appear not yet to have been investigated in relation to noise. While IgE is implicated in a variety of acute allergic phenomena, the other immunoglobulins appear to be linked to the body's ability to withstand infectious diseases, as well as to other important physiological and pathological processes. Clearly there is a need to learn more about the effects of noise on the immune function in all its aspects.

Neoplastic Effects

Although theirs is the only such report we were able to identify, the case-control study of acoustic neuromas by Preston-Martin and his colleagues is intriguing (1989). They assessed noise exposure on the basis of questionnaire responses, which were confirmed by a blinded review of job history by an occupational health specialist. The odds ratio (OR) for this neoplasm increased in noise-exposed persons (OR=2.2, 95% confidence interval[CI]=1.12-4.67). For those with over 20 years of noise exposure, the risk was much greater (OR=13.2, CI=2.01-86.98). Those with acoustic neuromas included painters, mechanics, or truck drivers who recalled prior exposure to solvents in the 1940s, whereas control subjects were so employed only in the recent past, and had very little solvent contact. The findings of Preston-Martin et al. (1989) warrant further investigation, despite possible difficulties such as recall bias, and the lack of direct measurements of exposure to solvents, radiation or other specific environmental factors.

Miscellaneous Effects

Noise may have other important effects, besides the above, on human performance or behavior that are not limited to a specific anatomic or physiologic realm. For example, Cherek (1985) found an increase in cigarette smoking that was linearly related to the loudness of recorded industrial noise of 70, 80 and 90 dBA intensity. He compared the number of cigarettes smoked in 2-hour laboratory exposures of resting subjects, using each as his own control with a 60 dBA background session to establish a baseline.

Adverse effects on safety are probably multifactorial in origin, but Broadbent (1980) has cited data that link noise exposure to injury rate, and noted that the use of ear protectors was followed by a drop in injuries in both noisy and quiet places. While one would wish to make appropriate allowances for the possible operation of a Hawthorne effect, the result is clearly welcome to managers, labor leaders, and regulatory officials alike. The role of distraction in job-related injuries has been suggested by van Dijk, Ettema and Zielhuis (1987), who reason that noise annoyance may cause workers to change their work strategies and neglect routine safety practices. The role of stress will remain unclear in the absence of an accepted definition of the term in most discussions of specific noise-induced outcomes. Surprisingly little field research has been published on this topic.

Noise-induced stress has also been identified as playing a role in burnout of critical-care nurses (Topf and Dillon 1988). Even though the sustained noise levels in intensive-care units rarely reach those associated with hearing loss, they include alarms and other urgent signals often requiring instantaneous responses of a highly complex nature. Steps to remedy this aspect of occupational stress among nurses deserve further attention.

Hypothesizing along related lines, other researchers have discussed "noise sickness" as a syndrome *preceding* actual loss of hearing and embodying many stress-related somatic complaints in the auditory, autonomic, cardiovascular, endocrine, and gastrointestinal systems (Westman and Walters 1981). This construct has heuristic value, but has not achieved wide acceptance.

Another area of prospective research is suggested by the report of Knipschild, Meyer, and Salle (1981) that aircraft noise may reduce birthweight. Whether the observed reductions were a direct effect, or one due to a shortening of the gestation period is not clear.

No confirmation of this noise effect was found by Dennler, Diener and Muller (1989), who also observed no impact on either spontaneous abortion rate or maternal well-being among 317 gravid women working in noise levels of 85 dBA compared to 489 pregnant women serving as controls. Other adverse reproductive effects of noise, such as the sort of fetal malformations seen in small mammals, were not found by Kuppka and associates (1989) in a study of Finnish women working amid levels of 80 dB or higher in their first trimester.

Finally, an additional area of physiology in which no significant information appears to have been published, but that might be a fruitful subject of investigation, is that of bronchopulmonary function in relation to noise exposure. Given the multitude of factors that have been shown to influence bronchomotor tone, it seems likely that some normal as well as some asthmatic individuals may exhibit increases in airway resistance in high-noise environments. It seems less likely that full-blown asthmatic reactions would be so induced, since the chances appear remote that such an important effect would have escaped detection. The effects of noise on such pulmonary functions as gas exchange and mucociliary clearance do not seem to afford promising opportunities for research, given the current state of our understanding of the mechanisms at work on this level.

REFERENCES

Abel, S.M. 1990. The extra-auditory effects of noise and annoyance: an overview of research. *The Journal of Otolaryngology,* Supplement 1:1-13.

Andrén, L., Hansson, L., and Björkman, M. 1981. Haemodynamic effects of noise exposure before and after beta 1-selective and non-selective beta-andrenoceptor blockade in patients with essential hypertension. *Clinical Science* (Supplement) 61:895-915.

Arguelles, A. E., Ibeas, D., Moses Ottone, J., and Chekherdemian, M. 1962. Pituitary-adrenal stimulation by sound of different frequencies. *Journal of Clinical Endocrinology* 22:846-852.

Arguelles, A. E., Martinez, M. A., Pucciarelli, E., and Disisto, M. V. 1970. In *Physiological Effects of Noise,* eds. B. L. Welch and A. M. S. Welch. New York: Plenum Press.

Axelsson, A., and Lindgren, F. 1980. Is there a relationship between hypercholesterolemia and noise-induced hearing loss, *Acta Otolaryngologica* (Stockholm) 100:379-386.

Babisch, W., Ising, H., Gallacher, J. E. J., and Elwood, P. C. 1988. Traffic noise and cardiovascular risk. The Caerphilly study, first phase. Outdoor noise levels and risk factors. *Archives of Environmental Health* 43:407-414.

Brandenberger, G., Follenius, M., Wittersheim, G., and Salamé, P. 1980. Plasma catecholamines and pituitary adrenal hormones related to mental task demand under quiet and noise conditions. *Biological Psychology* 10:239-252.

Broadbent, D. E. 1980. Noise in relation to annoyance, performance and mental health. *Journal of the Acoustical Society of America* 68(1):15-17.

Cantrell, R. W. 1974. Prolonged exposure to intermittent noise: audiometric, biochemical, motor, psychological and sleep effects. *Laryngoscope* 84 (Supplement 1):1-55.

Catapano, F., Portaleone, P., Teagno, P. S., Fornaca, G. F., Bono, F., Giuliani, G. C., Liberali, L., and Verdon di Cantogno, L. 1984. Effects of traffic noise on plasma catecholamines, cAMP and some metabolic and cardiovascular functions in a group of normal subjects. *Minerva Medica* (Italy) 75:1111-1115.

Cavatorta, A., Falzoi, M., Romanelli, A., Cigala, F., Ricco, M., Bruschi, G., Franchini, I., and Borghetti, A. 1987. Adrenal response in the pathogenesis

of arterial hypertension in workers exposed to high noise levels. *Journal of Hypertension* (Supplement 5) 5:S463-S466.
CESANA, G. C., FERRARIO, M., CURTI, R., ZANETTINI, R., GRIECO, A., SEGA, R., PALERMO, A., MARA, G., LIBRETTI, A., and ALGERI, S. 1982a. Work-stress and urinary catecholamines excretion in shift workers exposed to noise I: Epinephrine (E) and Norepinephrine (NE). *La Medicina del Lavoro* 2:99-109.
CESANA, G. C., PANZA, G., FERRARIO, M., CURTI, R., ZANETTINI, R., GRIECO, A., SEGA, R., PALERMO, A., MARA, G., LIBRETTI, A. and ALGERI, S. 1982b. Work-stress and urinary catecholamines excretion in shift workers exposed to noise II: Dopamine. *La Medicina del Lavoro* 2:110-117.
CHEREK, D. R. 1985. Effects of acute exposure to increased levels of background industrial noise on cigarette smoking behavior. *International Archives of Occupational and Environmental Health* 56:23-30.
COHEN, S., GLASS, D. C., and PHILLIPS, S. 1979. Environmental factors in health. In *Handbook of Medical Sociology*, eds. H.E. Freeman, S. Levine, and L.G. Reeder. Englewood Cliffs, N.J.: Prentice-Hall.
DENNLER, G., DIENER, L., and MULLER, W. 1989. Effects of noise on the foetoplacental unit (epidemiologic and experimental investigations). *Zeitschrift für die Gesamte Hygiene und ihre Grenzgebiete* 35:712-714.
ENGELAND, W. C., MILLER, P., and GANN, D. S. 1990. Pituitary-adrenal and adrenomedullary responses to noise in awake dogs. *American Journal of Physiology* 258 (Regulatory Integrative Comparative Physiology 27):R672-R677.
FOURIAUD, C., JACQUINET-SALORD, M. C., DEGOULET, P., AIMÉ, F., LANG, T., LAPRUGNE, J., MAIN, J., OECONOMOS, J., PHALENTE, J., and PRADES, A. 1984. Influence of socioprofessional conditions on blood pressure levels and hypertension control. Epidemiologic study of 6,665 subjects in the Paris district. *American Journal of Epidemiology* 120(6):72-86.
FROLICH, E. D. 1988. Recommendations for blood pressure determination by sphygmomanometry. *Annals of Internal Medicine* 109:612.
FRUHSTORFER, B., FRUHSTORFER, H., GRASS, P., MILERSKI, H. G., STURM, G., WESEMANN, W., and WIESEL, D. 1985. Daytime noise stress and subsequent night sleep: Interference with sleep patterns, endocrine and neurocrine functions. *International Journal of Neuroscience* 26:301-310.
GUILIAN, E. 1974. Noise as an occupational hazard: effects on performance level and health—a survey of findings in the European literature. National Institute for Occupational Safety and Health, Centers for Disease Control, U.S. Department of Health, Education and Welfare, Cincinnati.
HORVATH, S. M., and BEDI, J. F. 1990. Heat, cold, noise, and vibration. In *Environmental Medicine*, ed. A.C. Upton. The Medical Clinics of North America 74(2):515-525.
ISING, H., DIENEL, D., GUNTHER, T., and MARKERT, B. 1980. Health effects of traffic noise. *International Archives of Occupational and Environmental Health* 47:179-190.
JONDERKO, G., GABRYEL, A., JONDERKO, K., KOŃCA, A., MARCISZ, C., NEUMANN, M., OLAK, Z., and WYBRANIEC-PATALONG, A. 1982. The effects of physical effort, noise and vibration on thyroid function. *Endokrynologia Polska* 33:121-127.
KAUS, L. C., and FELL, J. T. 1984. Effect of stress on the gastric emptying of capsules. *Journal of Clinical and Hospital Pharmacy* 9:249-251.

KLOTZBÜCHER, E. 1976. Zum Einfluss des Lärms auf Leistung bei geistiger Arbeit und ausgewählte physiologische Funktionen. *International Archives of Occupational and Environmental Health* 37:139-155.

KLOTZBÜCHER, E., and FICHTEL, K. 1979. Zum Einfluss des Lärms auf die optische Signalerkennung. *Ergonomics* 22:919-926.

KNIPSCHILD, P. 1977. VI. Medical effects of aircraft noise: General practice survey. *International Archives of Occupational and Environmental Health* 40:191-196.

KNIPSCHILD, P., and OUDSHOORN, N. 1977. VII. Medical effects of aircraft noise: Drug survey. *International Archives of Occupational Archives of Occupational and Environmental Health* 40:197-200.

KNIPSCHILD, P., MEYER, H. and SALLE, H. 1981. Aircraft noise and birth weight. *International Archives of Occupational and Environmental Health* 48:131-136.

KUPPA, K., RANTALA, K., NURMINEN, T., HOLMBERG, P. C., and STARCK, J. 1989. Noise exposure during pregnancy and selected structural malformations in infants. *Scandinavian Journal of Work and Environmental Health* 15:111-116.

LISIEWICZ, J., and MOSZCZYNSKI, P. 1986. Effects of noise on the hematopoietic and immune systems II. Erythocytes, neutrophils, lymphocytes, immunoglobulins, biochemical changes in the blood and the hemostatic system. *Przeglad Lekarski* 43(3):331-336.

LOEB, M. 1986. *Noise and Human Efficiency*. New York: John Wiley & Sons.

MARTH, E., GALLASCH, E., FUEGER, G. F., and MÖSE, J. R. 1988. Aircraft noise: changes of biochemical parameters. *Zentralblatt für Bakteriologie, Mikrobiologie und Hygiene* 185:498-508.

OSGUTHORPE, J. D., and MILLS, J. H. 1982. Non-auditory effects of low-frequency noise exposure in humans. *Otolaryngology, Head and Neck Surgery* 90:367-370.

PETIOT, J. C., PARROT, P., LOBREAU, J. P. and SMOLIK, H. J. 1988. Cardiovascular responses to intermittent noise in Type A and Type B female subjects. *International Journal of Psychophysiology* 6:111-123.

PFEIFFER, C. J. 1968. Gastroenterologic aspects of manned space flight: Comments on gastrointestinal gas and environmental aspects of manned spaceflight. *Annals of the New York Academy of Sciences* 150:40-48.

PRESTON-MARTIN, S., THOMAS, D. C., WRIGHT, W. E., and Henderson, B. E. 1989. Noise trauma in the aetiology of acoustic neuromas in men in Los Angeles County, 1978-1985. *British Journal of Cancer* 59:783-786.

RESNEKOV, L. 1981. Noïse, radio-frequency radiation and the cardiovascular system. *Circulation* 63:264A-266A.

ROSENMAN, K. D. 1990. Environmentally related disorders of the cardiovascular system. In *Environmental Medicine*, ed. A. C. Upton. The Medical Clinics of North America 74(2):361-375.

SONNENBERG, A., DONGA, M., ERKENBRECHT, J. F., and WIENBECK, M. 1984. The effect of mental stress induced by noise on gastric acid secretion and mucosal blood flow. *Scandinavian Journal of Gastroenterology 19* (Supplement 89):45-48.

SUCZEWSKI, E. J. 1989. Personal communication.

TALBOTT, E., HELMKAMP, J., MATTHEWS, K., KULLER, L., COTTINGTON, E., and REDMOND, G. 1985. Occupational noise exposure, noise-induced hearing loss, and the epidemiology of high blood pressure. *American Journal of Epidemiology* 121(4):501-514.

TOPF, M., and DILLON, E. 1988. Noise-induced stress as a predictor of burnout in critical care nurses. *Heart and Lung* 17:567-574.

VAN DIJK, F. J. H., 1986. Non-auditory effects of noise in industry II. A review of the literature. *International Archives of Occupational and Environmental Health* 58:325-332.

VAN DIJK, F. J. H., ETTEMA, J. H., and ZIELHUIS, R. L. 1986. Non-auditory effects of noise in industry I. Introduction and study objectives. *International Archives of Occupational and Environmental Health* 58:321-323.

VAN DIJK, F. J. H., ETTEMA, J. H., and ZIELHUIS, R. L. 1987. Non-auditory effects of noise in industry VII. Evaluation, conclusions and recommendations. *International Archives of Occupational and Environmental Health* 59:147-152.

VAN DIJK, F. J. H., SOUMAN, A., and DE VRIES, F. 1983. Industrial noise, annoyance and blood pressure. In *Noise as a Public Health Problem. Proceedings of the 4th International Congress* (Turin) Vol. 1:615-627.

WESTMAN, J. C., and WALTERS, J. R. 1981. Noise and stress: a comprehensive approach. *Environmental Health Perspectives* 41:291-309.

WITTERSHEIM, G., BRANDENBERGER, G., and FOLLENIUS, M. 1985. Mental task-induced strain and its after-effect assessed through variations in plasma cortisol levels. *Biological Psychology* 21:123-132.

YOUNG, L. D., RICHTER, J. E., ANDERSON, K. O., BRADLEY, L. A., KATZ, P. O., MCELVEEN, L., OBRECHT, W. F., DALTON, C., and MILLER SNYDER, R. 1987. The effects of psychological and environmental stressors on peristaltic esophageal contractions in healthy volunteers. *Psychophysiology* 24:132-141.

The Effects of Noise on Sleep

CHARLES P. POLLAK

Noise has been described as the most ubiquitous pollutant of the industrial world (Beardwood 1982). This is true in both a spatial and temporal sense. Noise is an obtrusive feature of many locales, occurring near urban roads, rural arterial highways, airports, railways, industrial plants, and inside hospitals and apartments. It is even present in the bedrooms of those who snore. Noise is also present around the clock. All of the locales just mentioned have been found to produce excessive levels of noise at night. This is important because (1) noises are more annoying when they occur at times when people expect to rest or sleep, (2) noise can interrupt sleep, and (3) noise can also have subtle effects on sleep and autonomic functions that are detectable only with specialized instruments. The sleep effects of noise will be reviewed in what follows, both to see what general conclusions can be reached and to aid in the formulation of plans for future research.

Subjective Responses to Nocturnal Noise

Subjective responses to noise, including self-assessed amount and quality of sleep, have been recorded under both field and laboratory conditions. Such responses are correlated with objective laboratory measures of sleep, but large biases occur that are usually in the direction of underestimating sleep amount and quality. Subjective reports are often the only practical way of conducting large surveys, however. Instruments cannot measure the kind of information that they can provide, such as the degree of displeasure or annoyance induced by noise, a problem that has furnished a motive for the measures taken to date to confront the problems of noise, including the research on noise summarized here.

A 1,000-night joint European field study of noise effects concluded that subjective sleep quality decreased after noisy nights (Jurriens et al. 1983), and other studies show that subjective assessments of poor sleep quality appear to be related to the level of noise (Griefahn and Muzet 1978; Öhrström and Björkman 1983). Noise effects are also related to additional factors, including subjective noise susceptibility, age, gender (women may be more susceptible), impaired health, frequency of complaints (i.e., tendency to complain), lower socio-economic status, and use of hypnotic drugs (Langdon and Buller 1977; Gros, Griefahn, and Lang 1983). Indeed, sleep disturbances appear to be the most widespread of the distressing effects of noise (Kryter 1985; Gloag 1980), and noise is the most frequently cited environmental stressor upon sleep (Webb and Cartwright 1978).

Even when they are not sleeping or attempting to sleep, people expect not to be disturbed at night. When residents of the vicinities of five Australian airports were asked which three-hour period of the day they would most like to have free from aircraft noise, their first choice by far was 6-9 P.M., followed by 9-12 P.M., and 12-3 A.M., in that order (Bullen and Hede 1983).

Current standards of noise measurement, therefore, make special allowance for the occurrence of noise at night. The day-night average sound level (L_{dn} or DNL) recommended by the U.S. Environmental Protection Agency (EPA 1974) applies a penalty of 10 dB to events occurring between 10 P.M. to 7 A.M. L_{dn} values have been shown to correlate closely with annoyance of communities world-wide (Schultz 1978) and have become widely recommended as a descriptor for purposes of community noise assessment and land-use planning. The process by which the 10 dB value was determined may have been flawed, however (Shepherd 1987), and a smaller weighting has been recommended (Bullen and Hede 1983).

Prevalence of Noise-related Sleeping Problems

The prevalence of sleep disturbances in the vicinity of noise sources may be high. In a large, age-specific (41-43 years) sample of the residents of Amsterdam, 6.8% reported difficulty sleeping because of traffic noise (Meijer, Knipschild, and Salle 1985). Almost 60% of residents living about one mile from JFK airport in New York City reported sleep disturbance that decreased with greater distance from the airport (Borsky 1976).

It is remarkable that nearly half of the nearer residents surveyed in the latter study slept "very or extremely well," a finding that demonstrates the existence of one or more noise-susceptible subsets among the residents. Variables conferring increased noise susceptibility could include age, female sex, and low socioeconomic status (SES). SES tends to be lower near sources of noise such as airports, a circumstance that may explain some apparent noise effects, as suggested by Kryter (1980), but there is also evidence that noise sensitivity is greater among people of higher SES (Meijer, Knipschild, and Salle 1985).

Note that in descriptive studies of the population, these variables or others, singly or in combination, could explain sleep disturbance by themselves, even though they may happen to be correlated with noise levels. Another limitation of most such studies is the failure to consider additional, simultaneous sources of annoyance. When this had been done, subjective sleep quality was correlated with the presence of multiple pollutants (e.g., noise, smells, and dust) to an approximately equivalent degree (Gros, Griefahn, and Lang 1983).

Normal Sleep

Sleep is a naturally occurring state of reversible behavioral quiescence and reduced reactivity. When recording electrodes are applied to a sleeping person to detect brain waves, eye movements, and muscle activity, a complex series of events is revealed. The night is punctuated by distinct periods of heightened brain activity, darting rapid eye movements (REM), muscular twitches, and fluctuating blood pressure and pulse rate. Breathing varies in rate and depth and may even cease repeatedly. Such states are termed "REM sleep" and occur throughout the night at 90-100 minute intervals. NonREM sleep is conventionally divided into four stages. Stages three and four are the deepest, and are often termed "slow-wave sleep" on the basis of the appearance of their corresponding electroencephalograms (EEGs). These forms of sleep are concentrated in the first several hours of sleep. REM sleep is also relatively deep, and predominates during the later hours of sleep.

While sleep can be described in exquisite detail using electrophysiological recordings, the functions of both REM and nonREM sleep remain obscure. Disruption of the normal pattern of sleep causes impaired cognitive and motor functions and a feeling of sleepiness, but there is no convincing evidence that one form of sleep is needed more than the other or produces greater benefits during the waking state.

Arousal from Sleep

The auditory system is the chief means by which human beings monitor the environment during sleep. The immediate response to a sufficiently strong auditory stimulus is arousal, and it is this response that distinguishes sleep from pathological states of the central nervous system such as stupor or coma. Brief arousals (of less than several minutes) may not cause an awareness of their occurrence, and if they are sufficiently frequent, such unrecalled arousals may induce profound daytime sleepiness. This effect has been shown both experimentally (Bonnet 1985) and by clinical experience with sleep disorders, especially sleep apnea.

Stimuli occurring during sleep may also induce responses that fall short of full arousal. The central nervous system (CNS) is capable of complex discriminations among environmental stimuli during sleep, and exhibits graded responses, ranging from subtle alterations of the EEG to full arousal (Keefe, Johnson, and Hunter 1971).

Such responses are mediated by a brainstem mechanism, the reticular activating system (RAS) (Cohen 1977). This system distributes signals throughout the CNS, enabling rapid mobilization of cognitive, autonomic motor, and neuroendocrine systems to prepare the organism for fight or flight. An important property of this system is *habituation,* or the progressive diminution in the degree of response to repeated stimuli. If this did not occur, costly behaviors could be triggered in response to events that are not likely to be significant.

These properties of the RAS imply that the effects of environmental disturbances such as noise are multidetermined and complex; properties of both the stimulus and the organism must be considered. Important properties of the stimulus include intensity, temporal pattern, information content, and context. Relevant properties of the organism include principally the *meaning* of the sound to the individual. Meaning is derived from both simple processes such as habituation, and complex ones that connect a specific sound to significant past experiences (conditioning). An often-cited example is the ease with which arousal is produced by saying the sleeper's name rather than a meaningless name (Oswald, Taylor, and Treisman 1960; Williams, Morlock, and Morlock 1960; Zung and Wilson 1961; Langford, Meddis, and Pearson 1974). Many meaningful associations are specific to the individual and cannot be predicted in planning studies. Sizable inter-individual differences are therefore to be expected in most studies of noise and sleep.

Physiological Effects of Nocturnal Noise

Noise may cause delay of sleep onset, awakenings from sleep, shifts of EEG sleep stage, delayed return to sleep, or premature awakening. Noise may also produce cardiovascular responses during sleep. These effects usually occur promptly, but delayed nocturnal effects of diurnal noise have also been found.

Noise Stimuli. Earlier research was concerned in large part with aircraft sounds such as sonic booms, but with the realization that the most important source of noise in urban society is road traffic, studies have increasingly been conducted with subjects sleeping in their usual home environment where they are exposed mainly to traffic noise (Vallet 1979; Jurriens 1981; Wilkinson and Campbell 1984; Eberhardt and Axelsson 1987). Such in situ studies have also examined train noise (Vernet 1979) and building noises (such as TV or shouting, etc.) (Öhrström 1983; Rabinowitz et al. 1988). Noise levels have been experimentally varied by opening windows, double glazing windows, using earplugs, or moving subjects to quieter bedrooms (Jurriens et al. 1983).

Most sleep studies conducted in laboratories have used recorded sounds of aircraft or road traffic (Lukas 1977; Griefahn 1980; Muzet 1983). Recorded military noise (cannon shots) has recently been used (Griefahn 1988).

Sleep Responses. Various techniques have been used to assess sleep responses (Muzet 1983). There has been a trend away from laboratory studies to studies in the home. This has sometimes required a reduction of the range, sensitivity, or precision of measurements, but the challenge of in situ recording has also led to the use of more sophisticated techniques (Vernet 1979).

Two major classes of sleep responses have usually been recorded: upward sleep stage shifts and awakenings. When responses to sounds of increasing intensity are measured, changes in the sleep EEG are the first to appear (Keefe, Johnson, and Hunter 1971). They can be detected in all stages of sleep in response to tones far below (25-30 dB) the arousal threshold (the level required to produce an EEG characteristic of the waking state with or without a behavioral response). Noise peaks are associated with transient increases in EEG alpha and delta frequencies. Rapid eye movements decrease briefly, then increase sufficiently to increase the whole-night REM sleep amount (Wilkinson and Allison 1983). Cardiovascular responses are detectable only at higher sound levels (5-20 dB below the arousal threshold). These responses include heart rate acceleration and decreased finger pulse amplitude. Electrodermal responses (skin potential and resistance), as well as respiratory and motor responses, do not occur until the arousal threshold is reached.

Awakenings have been detected by subject report (Öhrström and Rylander 1982; Horonjeff et al. 1982), by the occurrence of bed movements (Öhrström and Rylander 1982), and by polygraphic sleep recordings. The latter are usually interpreted by visual pattern recognition (Rechtschaffen and Kales 1968), but computer-assisted analytical methods have recently begun to be used (Campbell and Wilkinson 1981; Jurriens et al. 1983).

The probability that a response will occur is related to several features of the stimulus: peak level, mean equivalent level (L_{eq}), temporal pattern (whether constant or impulsive), duration, and meaning. Arousals induced by laboratory traffic noise and detected by bed movements are more closely related to peak noise levels than to total noise energy (Öhrström and Björkman 1983). Summarizing data from studies employing sonic booms and traffic noise, Thiessen (1978, 1983) showed that the probability of a change in sleep increased in approximately linear fashion as peak noise intensities increased from about 30 dB to 80 dB. For a mixed-aged group of subjects, the probability of being awakened by the noise of a passing truck increased from about 18% for peak noise levels of 50 dB to about 42% for 70 dB (Thiessen 1978). Shifts of sleep without arousal were even more likely to occur at the same levels: about 38% for 50 dB and 73% for 70 dB. Using a measure of noise stimulation that took into account both its spectral characteristics and its duration—the "effective perceived noise level" (EPNL)—Lukas (1975) found that 50% of subjects were awakened by aircraft noise of 90 EPNdB and 100% were awakened by 120 EPNdB. In a later report, Lukas (1977) showed that the probability of awakening to a peak noise level of 50 dBA is 5%, and of 70 dBA is 30%. From a review of the literature, Griefahn (1980) concluded that a response consisting of a shift of at least one sleep stage (e.g., a shift from stage three to either stage two, to stage one, or to wakefulness) to a maximum sound level of 68 dBA could be expected in one-third of the population, of whom in turn one-third would be awakened. The occurrence of sub-arousal responses of sleep to noise suggests that noise-induced sleep problems may be more widespread than surveys based on self-reporting would suggest (Gloag 1980).

With an increasing number of sound stimuli per night, the number of sleep stage shifts increases linearly, while the number of awakenings increases at a decreasing rate, reaching a maximum of about 3.5 awakenings for 35 stimuli per night (Vernet 1979; Griefahn 1980). Higher levels of road noise are associated with poorer sleep, and intermittent noises have a greater effect than continuous ones (Öhrström and Rylander 1982).

Sleep in the presence of noise has also been quantified by time spent in various stages. In a review of studies conducted before 1980, Griefahn (1980) found only limited changes: waking time is slightly greater, time spent in slow-wave sleep is slightly reduced, but total sleep time and time spent in REM sleep or sleep stages one and two are unchanged. More recently, a joint European field study concluded that slightly less time was spent in REM sleep (mean difference, 6.5 minutes) during noisy nights (Jurriens et al. 1983).

Other whole-night measures of sleep have occasionally been employed, such as EEG power in the delta (0.5-2.5 Hz) spectral band, which decreases in the presence of taped traffic noise (Jurriens 1980; Wilkinson and Allison 1983). Analyses based on EEG spectral bands must be interpreted with caution, however. Transient responses may differ in sign from whole-night changes; thus, transient increases of EEG delta activity may be found on nights that have overall, noise-related decreases of delta activity (Wilkinson and Allison 1983). The apparent discrepancy is probably explained by the occurrence of noise-induced arousals that are associated with K-complexes and that interfere with the induction or maintenance of deep non-REM sleep stages.

Finally, while noise is often defined as unwanted sound, especially at night, it may also be useful, and even essential. Examples include calls or other signals by elderly or ill people to summon their caregivers at night, household smoke and fire detection devices, and the ordinary alarm clock used to terminate sleep. Studies have therefore been conducted to determine what stimulus is needed to induce arousal reliably. A 55 dBA smoke detector alarm signal was sufficient to awaken normal-hearing young adults under quiet background conditions. In the presence of background noise, 70-85 dbA signals were needed (Nober et al. 1980).

Cardiovascular Responses During Sleep. Cardiovascular responses to noise also occur during sleep. They consist of increases or biphasic changes of heart rate and changes in finger pulse amplitude caused by vasoconstriction (Keefe, Johnson, and Hunter 1971; Muzet and Ehrhart 1978; Wilkinson and Allison 1983). Positive correlation of mean heart rate and heart rate variability with sound level were found by the 1000-night joint European study (Jurriens et al. 1983). Such responses are actually greater during sleep than during wakefulness, even when the noise intensity is lower at night (Keefe, Johnson, and Hunter 1971; DiNisi and Muzet 1988). They are related to peak noise intensity and do not exhibit a habituation effect (Ehrenstein and Weber 1980; Muzet and Ehrhart 1980; Wilkinson and Allison 1983). Their significance for health is presently unknown, but if noise has long-term effects on health, they may be related to the lack of habituation of these responses (Muzet et al. 1983).

Effects of Diurnal Noise on Subsequent Sleep

Only noise occurring during the sleep period has been considered so far, but evidence has recently appeared that both ambient and laboratory diurnal noise can affect subsequent nocturnal sleep (Blois, Debilly, and Mouret 1980; Frühstorfer, Frühstorfer, and Grass 1984; Frühstorfer et al. 1985). Although the findings are not highly reproducible, they suggest that it may be important to measure the pattern of noise across the entire 24-hour day, and such studies are now beginning to be carried out (Frühstorfer 1988).

Habituation to Nocturnal Noise

If the effects of environmental noise on sleep lasted for only a part of the night or even a few nights, the environmental sources of such disturbances would be of limited social concern. Failure of sleep or cardiovascular responses to exhibit a full habituation effect, on the other hand, would raise concerns about effects on population groups that are chronically exposed to noise.

In accordance with the general principle that reticular activation is dependent upon stimulus significance, habituation of sleep responses has in fact been observed to occur with repetition of noise stimuli over a series of nights. According to the literature review in Griefahn 1980 (covering 8 publications, 72 subjects, and 8,138 noise stimuli), with repeated nightly noise stimulation, the number of awakenings decreases at a decreasing rate for about seven nights. Habituation specific to aircraft noise (as opposed to music of equivalent intensity and spectrum) can even develop in utero during the first five months of gestation (Ando and Hattori 1977).

Not all studies show habituation of sleep responses, however. Recorded military noise (cannon shots) of about 80 dBA produced changes in sleep stage that decreased over three hours of stimulus presentation, but did not exhibit habituation over 13 nights (Griefahn 1988). Both arousals signaled by nocturnal bed movements as well as heart rate responses to the acute effects of traffic noise did not habituate over two weeks (Öhrström 1988). These and other studies (Ehrenstein and Weber 1980) suggest that habituation may not develop for every type of noise.

For obvious reasons, it is not possible to expose research subjects to nightly noise for long periods of time such as residing in a noisy locale would involve. Evidence regarding long-term habituation therefore comes from two types of field studies. First, there are the studies already mentioned of sleep complaints from the residents of the environs of airports and highways. The methodological problems of such studies make it difficult to attribute the complaints to noise, however.

Additional evidence for incomplete habituation comes from field investigations in which the environment is modified to change the level of noise. Double glazing of windows, for example, has been found to improve sleep (Öhrström and Björkman 1983; Öhrström 1983) and even next-day performance (Wilkinson, Campbell, and Roberts 1980). The simplest interpretation of such data is that the improvement of sleep is explained by a reduction of noise and not by a Hawthorne effect (Roethlisberger and Dickson 1939), implying that the disruptive effects of noise on sleep had not fully habituated after years of residence.

Autonomic variables have also been studied in this regard. Adaptation of heart rate changes to traffic noise peaks has been found in healthy people living near busy arterial roads (Wilkinson and Allison 1983), but cardio-acceleratory responses to cannon shots did not decrease acutely or over 13 nights (Griefahn 1988), and autonomic responses have been found not to habituate even after five or more years of exposure to traffic noise (Vallet et al. 1988). Failure to obtain improvements in mood, performance, mean heart rate, and cardio-acceleratory responses with double glazing has been interpreted as evidence of insufficient acoustic efficacy of the double glazing (Kumar et al. 1983). It may also demonstrate that some effects of chronic sleep disturbance are not immediately reversed by noise reduction and that longer trials may be needed.

Special Groups

While noise is a public health problem relevant to all people, there may exist groups that are more susceptible to the effects of noise. If so, it would be important to identify these groups, explain their sensitivity to noise, and devise means of protecting them from its effects (Rehm and Gros 1980). Also to be considered is the level and nature of potential risks. Elderly people, for example, are exposed to the risk of falls and fractures whenever they attempt to get out of bed during the night when awakened by sounds.

The Elderly. On the strength of a review of two laboratory studies of sonic booms, Lukas (1975) concluded that the probability of arousal, whether determined by EEG criteria or observable as actual behavioral awakening, increased with age. Age-related differences in responses to truck noises have also been studied by Thiessen (1978). Whether behavioral awakenings or sleep stage shifts were considered, middle-aged subjects (46-51 years) were more likely to respond than either younger (16-25 years) or older subjects (55-77 years). Griefahn (1980), in a review study covering three publications, 26 subjects, and 4,428 stimuli, considered the influence of age on the probability that sonic booms and aircraft noises of 68-86 dBA would cause arousal. The number of responses increased with age from childhood to the mid-seventies. The amplitude of heart rate and finger pulse responses to noise during sleep is also age-dependent (Muzet et al. 1981). Most of the evidence, then, favors the view that noise-induced sleep disturbances increase monotonically with age.

There is also some contrary evidence. No age effects of continuous or impact (hammer-blow) noise on awakenings and sleep stage shifts were found by Roth, Kramer, and Trinder (1972), though slow-wave sleep was differentially reduced in the older subjects.

Survey data are very sparse. Although some elderly persons report sensitivity to neighborhood noises (Mant and Eyland 1988), several surveys have shown the elderly to be less likely to cite noise as a cause of sleep disturbance (Franke 1979; Rabinowitz et al. 1988). Since insomnia is considerably more frequent in the elderly (Karacan et al. 1976; Bixler et al. 1979), it is possible that noise-related sleep disturbances are simply masked by many additional causes of sleep disturbance.

The Young. The EEGs and finger plethysmograms of two- to four-month-old infants living in the vicinity of an airport were abnormal, despite evidence that they had developed habituation to aircraft noise (Ando and Hattori 1977).

Normal five- to seven-year-old children have higher arousal thresholds than adults and are less sensitive to aircraft noise. Normal infants are also relatively insensitive to noise, but premature or neurologically abnormal infants show less discriminate responses to sounds (Mills 1975). Traffic noises significantly disturb the sleep of prepubertal children (Eberhardt 1988).

The effects of environmental noise are illustrated by increased behavioral disturbances of sleep (such as longer sleep latency, more movements during sleep, and shorter sleep) observed in kindergarten children sleeping at school in a noisy area, contrasted to a quieter area (Havranek et al. 1979).

In view of the possible importance of sleep in brain development (Roffwarg, Muzio, and Dement 1966), somatic development, and recovery from daytime sensory and information loads (Horne 1979), the effects of noise on the sleep patterns of infants and children deserve further study.

Women and Men. There are few reliable data regarding the possibility that the sleep of women and men may be affected differently by noise. After reviewing the literature up to 1975, Lukas (1975) concluded that young women may be less responsive to nighttime noises than men, but women aged 35 and over are more responsive. A recent investigation of the effects of prerecorded cannon shots on polygraphically recorded sleep found no differences in men and women (Griefahn 1988). In a large, urban sample of 41- to 43-year-olds, the degree of annoyance and sleep disturbance attributed to noise did not differ in men and woman (Meijer, Knipschild, and Salle 1985). It thus remains an unresolved question whether or not consistent, age-adjusted differences in responses to noise exist in men and women.

The Sick. Noise in hospitals is both an annoyance and a potential threat to recovery from illness. The single greatest source of noise annoyance has been found to be loud talking in the hallway at night (Topf 1985).

It is commonly assumed that sound sleep is helpful to recovery from surgery or illness. Yet nocturnal noises of 40-45 dbA may be common in hospitals, and even louder noises in the daytime are likely to prevent patients from sleeping well (Aitken 1982). The recommended ceiling nighttime noise level for hospital wards is 35 dBA (EPA 1974). Noise measurements made in the early 1970s in the recovery room and acute-care unit of a large teaching hospital in the

United States showed noise levels during the night that were nearly as high as in the daytime—55 dBA from 11 P.M. to 7 A.M. and 58-59 dBA at other times (Falk and Woods 1973). In another study, levels greater than 50 dBA were present 25-80% of the time during the night on acute and general medical wards (Soutar and Wilson 1986; Bentley, Murphy, and Dudley 1977; Redding, Hargest, and Minsky 1977).

Continuous noise levels of 58 dBA have been recorded in nursery incubators. These levels are sufficient to interfere with sleep in both adults and infants (Gadeke et al. 1969; Williams 1970; Thiessen 1970).

Disruption of the 24-hour sleep-wake cycle often develops in intensive care units (ICUs) and has been correlated with the development of a temporary psychosis called "intensive care unit syndrome" (Hilton 1976; Helton, Gordon, and Nunnery 1980). The high levels of round-the-clock noise present in ICUs (Hilton 1976; Falk and Woods 1973) are often cited as being important, although additional factors are likely to be involved. The problem of ICU noise and noise control measures has also been reviewed from the nursing perspective (Baker 1984; Hansell 1984).

Shiftworkers. It is well known that night-shift workers are unable adequately to sustain daytime sleep. While the effort to sleep at an inappropriate time in the circadian sleep cycle is probably the main reason, it is also true that daytime conditions are usually not conducive to sleep. Noise, along with stress and use of coffee and tobacco, has been found to be predictive of the quality of daytime sleep (Frese and Harwick 1984).

Noise-Sensitive Poor Sleepers. Most studies of noise effects show considerable inter-individual variation. Some of this variation may be explained by a trait of noise sensitivity. A scale of noise sensitivity has been developed that predicts annoyance by dormitory noise (Weinstein 1978) and disturbance of postoperative patients by hospital noise (Topf 1985). Noise sensitivity may be very common. In a sample of 3,445 residents of Amsterdam, 28% regarded themselves as noise-sensitive, 37% as noise-insensitive, and 35% as in between (Meijer, Knipschild, and Salle 1985).

Persons who are annoyed by traffic noise also are annoyed by other environmental noises such as are emitted by airplanes, neighbors, and machinery at their places of work (Meijer, Knipschild, and Salle 1985)—evidence that noise sensitivity is an individual trait rather than a specific response to certain noises. It is possible that the sensitivity of such people is not limited to the auditory modality. They may be sensitive to a broad range of physical and social stimuli, both in the daytime and at night. Indeed, noise-sensitive individuals would seem to overlap with self-described poor sleepers, who also say their sleep is "light" and that they are easily awakened by noise

(Monroe 1967). Noise-sensitive individuals have only rarely formed separate groups in laboratory noise experiments (Öhrström 1988).

Considering the likelihood that environmental noise can induce lasting interference with sleep, little cognizance has been taken of this cause of chronic sleeping difficulties in the sleep literature. For example, despite increasing interest in insomnia in recent years, little is known about the noise exposure of insomniacs.

Risk factors may of course have additive or interactive effects. Thus, subgroups of the elderly with insomnia, other sleep disorders, depression, dementia, and so on, might be especially susceptible to the sleep-disturbing effects of noise. While there is some survey evidence that elderly people with sleep problems are sensitive to environmental noise (Mant and Eyland 1988), the matter remains largely uninvestigated.

Short-term Consequences of Noise-disturbed Sleep

Noise-induced disturbances of sleep can have substantial effects on subsequent task performance. The joint European study team mentioned earlier (Jurriens et al. 1983) concluded that simple reaction times are increased after noisy nights at home, and one of the constituent study teams also found increased errors on a four-choice reaction time test. Memory has been found to be impaired in some studies (LeVere and Bartus 1972), but not in others (Wilkinson, Campbell, and Roberts 1980). Multiple-choice reaction times were slowed after intermittent, but not continuous, traffic noise (Öhrström and Rylander 1982; Öhrström 1988), but are sometimes confounded by learning effects (Wilkinson, Campbell, and Roberts 1980). Impairments of mental arithmetic ability or pattern discrimination have not been found (Chiles and West 1972).

Another ill effect of nighttime noise is irritability in the morning (Öhrström 1983). Sleepiness would also be likely to develop, but this has not been investigated, as already noted.

Noise can also *facilitate* task performance, perhaps by the same arousal effects that interfere with sleep (Wilkinson 1963; Corcoran 1967). Whether or not performance decrements are found the day after a noisy night could, therefore, depend on whether or not the testing is carried out in the same noisy environment. This may have implications for the work environment, especially of shiftworkers.

The effects of noise on performance may not exhibit full habituation. Performance by Londoners on a simple reaction-time test taken in the morning improved when nighttime bedroom noise was reduced from over 50 dBA to 42 dBA by double glazing (Wilkinson, Campbell, and Roberts 1980).

Long-term Consequences: Implications for Health and Prevention of Illness

Research on noise since the 1950s has produced essentially no evidence that the effects of noise on sleep can have long-term biological consequences (Griefahn 1980; DeJoy 1984). A biological basis for such effects does appear to exist, however, since measured sleep effects, as well as abnormal cardiovascular responses and next-day performance decrements can persist for many years. In addition, there are persistent, but less well-defined effects of noise that are expressed as subjective sleep disturbance, and that can perhaps be measured as next-day performance deficits.

Whether or not sleep disruption induced by noise (or anything else) may have long-term health implications is not known. Recent epidemiological evidence has linked subjects' reports of difficulty in sleeping to institutionalization and longevity among the urban elderly (Pollak et al. 1990). Even correlative data relating round-the-clock noise exposure to long-term health effects has yet to be collected. The development of ambulatory monitoring techniques within the past several decades has made such studies technically feasible. The 24-hour pattern of noise exposure of urban and suburban factory workers has been calculated from measurements made at the places visited by them in the course of the day (von Gierke 1976). The biggest difference is at night, when the urban worker is exposed to about 52 dB (L_{eq}), and the suburban worker to 35 dB (L_{eq}).

A secondary effect that might have health implications is the use of tranquilizers and hypnotic drugs by people who are exposed to noise, such as those living near airports (Knipschild and Oudshoorn 1977). Intermediate- and long-acting benzodiazepine sedative-hypnotics have been shown to improve sleep induction and maintenance in noise-exposed subjects (Saletu et al. 1987; Saletu, Grunberger, and Sieghart 1985). Some of these effects may be explained by the ability of such drugs to raise the arousal threshold for noise (Bonnet, Webb, and Barnard 1979; Johnson et al. 1979; Johnson and Spinweber 1983), though arousal thresholds do not differ in good and poor sleepers (Johnson et al. 1979). Benzodiazepines have only limited effect on cardiovascular responses (Muzet 1983; Libert et al. 1988), however, and are therefore unlikely to abolish all long-term effects of environmental noise. Furthermore, they may carry risks of their own.

Conclusions

Further laboratory measurements of the effects of noise on sleep are needed. Representative subject samples are needed, focusing on noise-sensitive people and others who may be at increased risk. Intermittent noise is more disruptive than continuous noise, but descriptors of the temporal pattern of noise are lacking, and the characteristics of intermittent stimuli need to be standardized.

Noise research appears to have passed through an initial phase of descriptive investigations utilizing simple, univariate designs. It has been learned that responses to noise depend on factors in addition to the characteristics of the noise itself. These include age, perhaps sex, and noise sensitivity. Other factors may also deserve consideration: socioeconomic status, personality, preexisting sleep problems, and sleep disorders (such as sleep apnea or periodic movements) that independently disrupt the continuity of sleep. It seems important to include such factors in future surveys, in situ studies, and laboratory studies so that their individual and joint ability to enable us to predict responses to noise can be measured. It will also make it possible to identify subgroups of the population that are vulnerable to noise.

Multiple aspects of noise stimuli should also be considered. For example, the specificity of noise effects should be examined by including additional noxious environmental stimuli in study designs.

The short-term (next-day) effects of nocturnal noise exposure need further study. Both sleepiness and impaired performance should be studied over the entire waking day.

Habituation to the effects of noise is incomplete for sleep and next-day performance, and is practically nonexistent for cardiovascular responses. Noisy environments may have long-term effects on health and productivity, in addition to the known effects on well-being, that remain to be identified. Ambulatory monitoring of noise exposure coupled with long-term follow-up of subjects could be fruitful.

Practical and safe remedies for noise-related sleep disturbances remain to be developed. Candidate measures could include reducing noise exposure by double glazing of windows, relocation of sleeping areas, or change of employment; controlling noise by reducing its intensity or altering its temporal pattern; and blunting the effects of noise on sleep by treatment of co-morbid sleep disorders or, nonspecifically, by administration of hypnotic drugs. Interventions should be continued for several weeks before their worth is judged. All treatment trials as well as the social implementation of ameliorative measures would be facilitated by the identification of risk groups.

REFERENCES

AITKEN, R.J. 1982. Quantitative noise analysis in a modern hospital. *Archives of Environmental Health* 37:361-364.

ANDO, Y., and HATTORI, H. 1977. Effects of noise on sleep of babies. *Journal of the Acoustical Society of America* 62(1):199-204.

BAKER, C.F. 1984. Sensory overload and noise in the ICU: Sources of environmental stress. *Critical Care Quarterly* 6(4):66-80.
BEARDWOOD, C.J. 1982. Hormonal changes associated with auditory stimulation. In *Handbook of Psychiatry and Endocrinology,* eds. G.D. Burrows and P.J.V. Beaumont. Amsterdam: Elsevier Biomedical Press.
BENTLEY, S., MURPHY, F., and DUDLEY, H. 1977. Perceived noise in surgical wards and in intensive care areas: An objective analysis. *British Medical Journal* 2:1503-1506.
BIXLER, E.O., KALES, A., SOLDATOS, C.R., KALES, J.D., and HEALEY, S. 1979. Prevalence of sleep disorders in the Los Angeles metropolitan area. *American Journal of Psychiatry* 136:1257-1262.
BLOIS, R., DEBILLY, G., and MOURET, J. 1980. Daytime noise and its subsequent sleep effects. In *Proceedings of the Third International Congress on Noise as a Public Health Problem* (Freiburg), *ASHA Reports 10,* eds. J. Tobias, G. Jansen, and W.D. Ward. Rockville (Maryland): American Speech-Language-Hearing Association.
BONNET, M.H. 1985. Effect of sleep disruption on sleep, performance, and mood. *Sleep* 8:11-19.
BONNET, M.H., WEBB, W.B., and BARNARD, G., 1979. Effect of flurazepam, pentobarbital, and caffeine on arousal threshold. *Sleep* 1:271-279.
BORSKY, P.N. 1976. Sleep interference and annoyance by aircraft noise. *Sound and Vibration* 10:18-21.
BULLEN, R.B., and HEDE, A.J. 1983. Time-of-day corrections in measures of aircraft noise exposure. *Journal of the Acoustical Society of America* 73(5):1624-1630.
CAMPBELL, K.B., and WILKINSON, R.T. 1981. Sleep in the natural environment: Physiological and psychological recording and analyzing techniques. In *Biological Rhythms, Sleep and Shift Work,* ed. L.C. Johnson. New York: Spectrum.
CHILES, W.D., and WEST, P. 1972. Residual performance effects of simulated sonic booms introduced during sleep. *FAA Tech. Report 72-19.* Oklahoma City: Federal Aviation Administration.
COHEN, A. 1977. Extraauditory effects of acoustic stimulation. In *Handbook of Physiology* Section 9, ed. E.H.K. Lee. Bethesda: American Physiological Society.
CORCORAN, B.W.J. 1967. Noise and loss of sleep. *Quarterly Journal of Experimental Psychology* 14:178-182.
DEJOY, D.M. 1984. The nonauditory effects of noise: Review and perspectives for research. *Journal of Auditory Research* 24:123-150.
DINISI, J., and MUZET, A. 1988. Cardiovascular responses to noise during waking and sleeping. (Abstract). In *Noise as a Public Health Problem,* Vol. 1, ed. T. Lindvall. Stockholm: Swedish Council for Building Research.
EBERHARDT, J.L., and AXSELSSON, K.R. 1987. The disturbance by road traffic noise of the sleep of young male adults as recorded in the home. *Journal of Sound and Vibration* 114:417-434.
EBERHARDT, J.L. 1988. The disturbance of the sleep of prepubertal children by road traffic noise as studied in the home. In *Noise as a Public Health Problem,* Vol. 1, ed. T. Lindvall. Stockholm: Swedish Council for Building Research.
EHRENSTEIN, W., and WEBER, F. 1980. The effect of having noise presented during 8 consecutive days on sleep stage patterns, mood and vegetative

functions. In *Sleep 1980. 5th European Congress on Sleep Research,* ed. L. Popoviciu. Basel: S. Karger.

EPA (U.S. Environmental Protection Agency) 1974. *Information on levels of environmental noise requisite to protect public health and welfare with an adequate margin of safety.* EPA 550/9-74-004. Washington, D.C.

Falk, S.A., and Woods, N.F. 1973. Hospital noise-levels and potential health hazards. *New England Journal of Medicine* 289:774-781.

Franke, H. 1979. Über das physiologische und pathologische Schlaf- und Wachverhalten von betagten. *Zeitschrift für Gerontologie* 12:187-199.

Frese, M., and Harwich, C. 1984. Shiftwork and the length and quality of sleep. *Journal of Occupational Medicine* 26:561-566.

Frühstorfer, B. 1988. Daytime noise load—a 24 hour problem? In *Noise as a Public Health Problem,* Vol. 1, ed. T. Lindvall. Stockholm: Swedish Council for Building Research.

Frühstorfer, B., Frühstorfer, H., and Grass, P. 1984. Daytime noise and subsequent night sleep in man. *European Journal of Applied Physiology* 53:159-163.

Frühstorfer, B., Frühstorfer, H., Grass, P., Milerski, H.G., Sturm, G., Wesemann, W., and Wiesel, D. 1985. Daytime noise stress and subsequent night sleep: Interference with sleep patterns, endocrine and neurocrine functions. *International Journal of Neuroscience* 26(3-4):301-310.

Gadeke, R., Doring, B., Keller, F., and Vogel, A. 1969. The noise level in a children's hospital and the wake-up threshold in infants. *Acta Pediatrica Scandinavica* 58:164-170.

Gloag, D. 1980. Noise and health: Public and private responsibility. *British Medical Journal* 281:1404-1406.

Griefahn, B. 1980. Research on noise-disturbed sleep since 1973. In *Proceedings of the Third International Congress on Noise as a Public Health Problem* (Freiburg), *ASHA Reports 10,* eds. J. Tobias, G. Jansen, and W.D. Ward. Rockville (Maryland): American Speech-Language-Hearing Association.

Griefahn, B. 1988. Effects of military noise during sleep. Relations to sex, time of night, and duration of presentation period. (Abstract). In *Noise as a Public Health Problem,* Vol. 1, ed. T. Lindvall. Stockholm: Swedish Council for Building Research.

Griefahn, B., and Muzet, A. 1978. Noise-induced sleep disturbances and their effects on health. *Journal of Sound and Vibration* 59:99-106.

Gros, E., Griefahn, B., and Lang, D., 1983. Sleep disturbances caused by noise: Analysis of a cross-sectional inquiry. In *Proceedings of the Fourth International Congress on Noise as a Public Health Problem* (Turin), Vol. 2, ed. G. Rossi. Milan: Edizioni Tecniche a cura del Centro Ricerche e Studi Amplifon.

Hansell, H.N. 1984. The behavioral effects of noise on man: The patient with "intensive care unit psychosis." *Heart and Lung* 13(1):59-65.

Havranek, J., Bartuskova, M., Zubikova, L., and Klominek, M. 1979. Untersuchungen zum Einfluss des Aussenlärmes auf den Nachmitagsschlaf von Kindern in Kindergarten. *Zeitschrift für die Gesamte Hygiene und ihre Grenzgebiete* (Berlin) 25:166-167.

Helton, M., Gordon, S., and Nunnery, S. 1980. The correlation between sleep deprivation and the intensive care unit syndrome. *Heart and Lung* 9:464-468.

Hilton, B.A. 1976. Quantity and quality of patients' sleep and sleep-disturbing factors in a respiratory intensive care unit. *Advances In Nursing* 1:453-468.

HORNE, J.A. 1979. Restitution and human sleep: A critical review. *Physiological Psychology* 7:115-125.

HORONJEFF, R.D., FIDELL, S., TEFFETELLER, S.R., and GREEN, D.M. 1982. Behavioral awakenings as functions of duration and detectability of noise intrusions in the home. *Journal of Sound and Vibration* 84:327-336.

JOHNSON, L.C., CHURCH, M.W., SEALES, D.M., and ROSSITER, V.S. 1979. Auditory arousal thresholds of good sleepers and poor sleepers with and without flurazepam. *Sleep* 1(3):259-270.

JOHNSON, L.C., and SPINWEBER, C.L. 1983. Benzodiazepine effects on arousal threshold during sleep. In *Proceedings of the Fourth International Congress on Noise as a Public Health Problem* (Turin), Vol. 2, ed. G. Rossi. Milan: Edizioni Tecniche a cura del Centro Ricerche e Studi Amplifon.

JURRIENS, A.A. 1980. Sleeping twenty nights with traffic noise: Results of laboratory experiments. In *Proceedings of the Third International Congress on Noise as a Public Health Problem* (Freiburg), *ASHA Reports 10,* eds. J. Tobias, G. Jansen, and W.D. Ward. Rockville (Maryland): American Speech-Language-Hearing Association.

JURRIENS, A.A. 1981. Noise and sleep in the home: Effects on sleep stages. In *Sleep 1980. 5th European Congress on Sleep Research* , ed. L. Popoviciu. Basel: S. Karger.

JURRIENS, A.A., GRIEFAHN, B., KUMAR, A., VALLET, M., and WILKINSON, R.T. 1983. An essay in European research collaboration: Common results from the project on traffic noise and sleep in the home. In *Proceedings of the Fourth International Congress on Noise as a Public Health Problem* (Turin), Vol. 2, ed. G. Rossi. Milan: Edizioni Tecniche a cura del Centro Ricerche e Studi Amplifon.

KARACAN, I., THORNBY, J., ANCH, M., HOLZER, C.E., WARHEIT, G.J., SCHWAB, J.J., and WILLIAMS, R.L. 1976. Prevalence of sleep disturbance in a primarily urban Florida county. *Social Science and Medicine* 10(5):239-244.

KEEFE, F.B., JOHNSON, L.C., and HUNTER, E.J. 1971. EEG and autonomic response pattern during waking and sleep stages. *Psychophysiology* 8(2):198-212.

KNIPSCHILD, P., and OUDSHOORN, N. 1977. Medical effects of aircraft noise: Drug survey. *International Archives of Occupational and Environmental Health* 40:197-200.

KRYTER, K.D. 1980. Physiological acoustics and health. *Journal of the Acoustical Society of America* 68(1):10-14.

KRYTER, K.D. 1985. *The Effects of Noise on Man.* 2d ed. Orlando: Academic Press.

KUMAR, A., TULEN, J.H.M., HOFMAN, W.F., VAN DIEST, R., and JURRIENS, A.A. 1983. Does double glazing reduce traffic noise disturbance during sleep? In *Proceedings of the Fourth International Congress on Noise as a Public Health Problem* (Turin), Vol. 2, ed. G. Rossi. Milan: Edizioni Tecniche a cura del Centro Ricerche e Studi Amplifon.

LANGDON, F.J., and BULLER, I.B. 1977. Road traffic noise and disturbance to sleep. *Journal of Sound and Vibration* 50:13-28.

LANGFORD, G.W., MEDDIS, R., and PEARSON, A.J.D. 1974. Awakening latency from sleep for meaningful and nonmeaningful stimuli. *Psychophysiology* 11:1-5.

LAVERE, R.T., and BARTUS, R.T. 1972. Electroencephalographic and behavioral effects of nocturnally occurring jet aircraft sounds. *Aerospace Medicine* 43:384-389.

LIBERT, J.P., AMOROS, C., MUZET, A., EHRHARD, J., and DINISI, J. 1988. Effects of triazolam on heart rate level and on phasic cardiac response to noise during sleep. *Psychopharmacology* 96:188-193.

LUKAS, J.S. 1975. Noise and sleep: A literature review and a proposed criterion for assessing effect. *Journal of the Acoustical Society of America* 58(6):1232-1242.

LUKAS, J.S. 1977. Measures of noise level: Their relative accuracy in predicting objective and subjective responses to noise during sleep. In EPA Report 600/1-77-010. Washington, D.C.: U.S. Environmental Protection Agency.

MANT, A., and EYLAND, E.A. 1988. Sleep patterns and problems in elderly general practice attenders: An Australian survey. *Community Health Studies* 12:192-199.

MEIJER, H., KNIPSCHILD, P., and SALLE, H. 1985. Road traffic noise annoyance in Amsterdam. *International Archives of Occupational and Environmental Health* 56(4):285-297.

MILLS, J.H. 1975. Noise and children: A review of the literature. *Journal of the Acoustical Society of America* 58(4):767-779.

MONROE, L.J. 1967. Psychological and physiological differences between good and poor sleepers. *Journal of Abnormal Psychology* 72:255-264.

MUZET, A. 1983. Research on noise-disturbed sleep since 1978. In *Proceedings of the Fourth International Congress on Noise as a Public Health Problem* (Turin), Vol. 2, ed. G. Rossi. Milan: Edizioni Tecniche a cura del Centro Ricerche e Studi Amplifon.

MUZET, A., and EHRHARD, J. 1978. Amplitude des modifications cardiovasculaires provoquées par le bruit au cours du sommeil. *Coeur et Médicine Interne* 17:49-56.

MUZET, A., and EHRHART, J. 1980. Habituation of heart rate and finger pulse responses to noise in sleep. In *Proceedings of the Third International Congress on Noise as a Public Health Problem* (Freiburg), *ASHA Reports 10,* eds. J. Tobias, G. Jansen, and W.D. Ward. Rockville (Maryland): American Speech-Language-Hearing Association.

MUZET, A., EHRHART, J., ESCHENLAUER, R., and LIENHARD, J.T. 1981. Habituation and age differences of cardiovascular responses to noise during sleep. In *Sleep 1980. 5th European Congress on Sleep Reasearch,* ed. L. Popoviciu. Basel: S. Karger.

MUZET, A., WEBER, L.D., AMOROS, C., EHRHART, J., and LIBERT, J.P. 1983. Electrophysiological and cardiovascular responses to noise during sleep. Effects of a benzodiazepine hypnotic. In *Proceedings of the Fourth International Congress on Noise as a Public Health Problem* (Turin), Vol. 2, ed. G. Rossi. Milan: Edizioni Tecniche a cura del Centro Ricerche e Studi Amplifon.

NOBER, E.H., PEIRCE, H., WELL, A., JONSON, C.C., and CLIFTON, C. 1980. *Waking Effectiveness of Household Smoke and Fire Detection Devices.* NBS-GCR-80-284. Springfield (Virginia): National Technical Information Service.

ÖRHSTRÖM, E. 1983. Sleep disturbances—aftereffects of different traffic noises. In *Proceedings of the Fourth International Congress on Noise as a Public Health Problem* (Turin), Vol. 2, ed. G. Rossi. Milan: Edizioni Tecniche a cura del Centro Ricerche e Studi Amplifon.

ÖHRSTRÖM, E. 1988. Primary and after effects on noise during sleep with reference to noise sensitivity and habituation-studies in laboratory and field. In *Noise as a Public Health Problem,* Vol. 1, ed. T. Lindvall. Stockholm: Swedish Council for Building Research.

Öhrström, E., and Rylander, R. 1982. Sleep disturbance effects of traffic noise: A laboratory study on aftereffects. *Journal of Sound and Vibration* 84:87-103.
Öhrström, E., and Björkman, M. 1983. Sleep disturbance before traffic noise attenuation in an apartment building. *Journal of the Acoustical Society of America* 73(3):877-879.
Oswald, I., Taylor, A.M., and Treisman, A. 1960. Discriminative responses to stimulation during human sleep. *Brain* 83:440-453.
Pollak, C.P., Perlick, D., Linsner, J.P., Wenston, J., and Hsieh, F. 1990. Sleep problems in the community elderly as predictors of death and nursing home placement. *Journal of Community Health* 15:123-135.
Rabinowitz, J., Bakonyi, M., Bockquet, J.J., Meyer, G.R., Olivetti, R., and Rey, P. 1988. Effects of noise in multi-storey buildings. (Abstract). In *Noise as a Public Health Problem,* Vol. 3, ed. T. Lindvall. Stockholm: Swedish Council for Building Research.
Rechtschaffen, A., and Kales, A. 1968. *A Manual of Standardized Terminology, Techniques and Scoring System for Sleep Stages of Human Subjects.* Washington, D.C.: U.S. Public Health Service, U.S. Government Printing Office.
Redding, J.S., Hargest, T.S., and Minsky, S.H. 1977. How noisy is intensive care? *Critical Care Medicine* 5(6):275-276.
Rehm, S., and Gros, E. 1980. Physiological effects of noise in critical groups. In *Proceedings of the Third International Congress on Noise as a Public Health Problem* (Freiburg), *ASHA Reports 10,* eds. J. Tobias, G. Jansen, and W.D. Ward. Rockville (Maryland): American Speech-Language-Hearing Association.
Roethlisberger, F.J., and Dickson, W.J. 1939. *Management and the Worker—An Account of a Research Program Conducted by the Western Electric Company, Hawthorne Works.* Cambridge: Harvard University Press.
Roffwarg, H.P., Muzio, J.N., and Dement, W.C. 1966. Ontogenetic development of the human sleep-dream cycle. *Science* 152:604-619.
Roth, T., Kramer, M., and Trinder, J. 1972. The effect of noise during sleep on the sleep patterns of different age groups. *Canadian Psychiatric Association Journal* 17:197-201.
Saletu, B., Grunberger, J., and Sieghart, W. 1985. Nocturnal traffic noise, sleep, and quality of awakening: Neurophysiologic, psychometric, and receptor activity changes after quazepam. *Clinical Neuropharmacology* (Supplement 1)8:S74-S90.
Saletu, B., Kindshofer, G., Anderer, P., and Grunberger, J. 1987. Short-term sleep laboratory studies with cinolazepam in situational insomnia induced by traffic noise. *International Journal of Clinical Pharmacology Research* 7(5):407-418.
Schultz, T.J. 1978. Synthesis of social surveys and noise annoyance. *Journal of the Acoustical Society of America* 64(2):377-406.
Shepherd, W.T. 1987. Annoyance characterization by noise metrics. In *Environmental Annoyance: Characterization, Measurement and Control,* ed. H.S. Koelega. Amsterdam: Elsevier Science Publishers B.V. (Biomedical Division).
Soutar, R.L., and Wilson, J.A. 1986. Does hospital noise disturb patients? *British Medical Journal [Clinical Research]* 292(6516):305.
Thiessen, G.J. 1970. Effects of noise during sleep. In *Physiological Effects of Noise,* ed. A.S. Welch. New York: Plenum Press.

THIESSEN, G.J. 1978. Disturbance of sleep by noise. *Journal of the Acoustical Society of America* 64(1):216-222.
THIESSEN, G.J. 1983. Effect of intermittent and continuous traffic noise on various sleep characteristics and their adaptation. In *Proceedings of the Fourth International Congress on Noise as a Public Health Problem* (Turin), Vol. 2, ed. G. Rossi. Milan: Edizioni Tecniche a cura del Centro Ricerche e Studi Amplifon.
TOPF, M. 1985. Personal and environmental predictors of patient disturbance due to hospital noise. *Journal of Applied Psychology* 70(1):22-28.
VALLET, M. 1979. Psychophysiological effects of exposure to aircraft or road traffic noise. *Proceedings of the Institute of Acoustics* (meeting), 3:1-4.
VALLET, M., LETISSERAND, D., OLIVIER, D., LAURENS, J.F., and CLAIRET, J.M. 1988. Effects of road traffic noise on the rate of heart beat during sleep. (Abstract). In *Noise as a Public Health Problem,* Vol. 1, ed. T. Lindvall. Stockholm: Swedish Council for Building Research.
VERNET, M. 1979. Effect of train noise on sleep for people living in homes bordering the railway line. *Journal of Sound and Vibration* 66:483-492.
VON GIERKE, H.E. 1976. Development of a uniform approach to characterize noise impact on people. *Aviation, Space, and Environmental Medicine* 47(1):45-53.
WEBB, W.B., and CARTWRIGHT, R.D. 1978. Sleep and dreams. *Annual Review of Psychology* 29:223-252.
WEINSTEIN, N. 1978. Individual differences in reactions to noise: A longitudinal study in a college dormitory. *Journal of Applied Psychology* 63:458-466.
WILKINSON, R.T. 1963. Interaction of noise with knowledge of results and sleep deprivation. *Journal of Experimental Psychology* 66:332-337.
WILKINSON, R.T., CAMPBELL, K.B., and ROBERTS, L.D. 1980. Effect of noise at night upon performance during the day. In *Proceedings of the Third International Congress on Noise as a Public Health Problem* (Freiburg), *ASHA Reports 10,* eds. J. Tobias, G. Jansen, and W.D. Ward. Rockville (Maryland): American Speech-Language-Hearing Association.
WILKINSON, R.T., and ALLISON, S. 1983. Effects of peaks of traffic noise during sleep on ECG and EEG. In *Proceedings of the Fourth International Congress on Noise as a Public Health Problem* (Turin), Vol. 2, ed. G. Rossi. Milan: Edizioni Tecniche a cura del Centro Ricerche e Studi Amplifon.
WILLIAMS, H.L. 1970. Auditory stimulation, sleep loss and the EEG stages of sleep. In *Physiological Effects of Noise,* ed. A.S. Welch. New York: Plenum Press.
WILLIAMS, H.L., MORLOCK, H.C., and MORLOCK, J.J. 1960. Instrumental behavior during sleep. *Brain* 83:440-453.
WILKINSON, T., and CAMPBELL, K.B. 1984. Effects of traffic noise on quality of sleep: Assessment by EEG, subjective report, or performance the next day. *Journal of the Acoustical Society of America* 75(2):468-75.
ZUNG, W.W.K., and WILSON, W.P. 1961. Response to auditory stimulation during sleep. *Archives of General Psychiatry* 4:548-552.

The Effects of Noise on Fetal Development

ANITA T. PIKUS

Noise and the Developing Fetus

Fetal Auditory Development. Increased knowledge of the normal embryological stages in the development of the human auditory system has enhanced the opportunity for the study of risks to the fetus. Exposure to noise is one of those risks. Normal maturation of responses to auditory stimuli seems to take place in the third trimester, and the human infant's auditory system is fully developed by term. According to the most recent studies, there remains no dispute as to whether or not noise actually reaches the fetal ear (Gerhardt et al. 1988; Kisilevsky, Muir, and Low 1989). According to Gerhardt (1990), the fetal acoustic environment can be described as a "low-pass filtered version" of the acoustic atmosphere surrounding the mother (a view endorsed in Vince et al. 1982; Gerhardt et al. 1988; and Querleu et al. 1988). Therefore, one obvious risk associated with maternal exposure to environmental noise would appear to be the possibility of damage to the peripheral auditory system of the developing infant's ear (Henry 1983; Lalande, Hétu, and Lambert 1986).

Fetal Susceptibility to Noise. On the evidence of studies with animal subjects across various species, Ryals (1990) has put forward the concept of a critical period of heightened susceptibility to noise-induced damage to the cochlea during its development. (The duration of this period varies widely as a function of species.) It has been hypothesized that such a period may likely exist during the final maturational stages of the hair cells and their innervation, but before the completion of the auditory nerve efferent system. Efferent fibers appear to be active in a cochlear protective response to acoustic trauma (Rajan and Johnstone 1988), and their absence at this earlier stage in the final development of the cochlea may account for the existence of a critical period of increased susceptibility to noise (Ryals 1990). The traumatic effects of noise are experienced by younger animals at lower exposure levels and at different frequencies than by adults (Henry 1983; Saunders, Dear, and Schneider 1985). Rubel and Ryals (1983) have suggested that there are changes in the frequency/place code in the cochlea during its development, and Ryals (1990) has proposed that the underlying processes of these changes may also be active during early increased susceptibility to noise. Post-mitotically-induced regeneration of hair cells following acoustic trauma has been demonstrated in animals (Corwin and Cotanche 1988; Ryals and Rubel 1988). This work has sparked general interest, and may be of importance to the determination of a critical period for acoustic trauma if the potential for hair cell regeneration is found to be related to maturational state within species (Ryals 1990).

No determination of such a critical period in the development of the human fetus has yet been made. If it does exist and is found to coincide with the final stages of cochlear maturation, it should occur roughly between 12 and 16 weeks before birth (Lavigne-Rebillard and Pujol 1988). Ryals (1990) points out that intrauterine attenuation should sufficiently protect the developing human fetal cochlea from excessive noise levels to which the mother may be exposed. However, the critical period of the later-developing human fetus could possibly extend into neonatal life, as it does for some other species (Ryals 1990). If this were the case for an individual premature baby, exposure to noise at the levels frequently encountered in the neonatal intensive care unit (NICU) could conceivably damage its still-developing cochlea. Follow-up studies of children exposed to incubator noises during early infancy have been largely inconclusive in terms of associated hearing loss except when the noise exposure was in synergy with ototoxicity. In one such study, incubated babies had more sensorineural hearing loss than non-incubated children (Henry 1983; Ryals 1990). Continued efforts to make incubator and other NICU noise quieter seem warranted, and further research is needed concerning the existence and nature of a possible human critical period for susceptibility to noise.

Fetal Startle Response Testing. There are numerous studies (for example, Kuhlman et al. 1988; Smith et al. 1988) to be found in the obstetrical literature investigating fetal startle responses to vibratory and acoustic stimulation supplied by the loud buzzing sound of an electronic artificial larynx (EAL) placed in contact with the mother's abdomen. It has been determined on the basis of neonatal auditory brainstem response (ABR) testing that hearing is not damaged by this procedure (Ohel et al. 1987). However, since sound pressure levels produced by the EAL can reach 135 dB in fluid, caution is still recommended with these procedures (Gerhardt 1990).

Fetal Monitoring with Ultrasound. For those concerned with fetal imaging as a means of monitoring fetal development, ultrasound has become a valuable tool. The safety of the procedure and its long-term effects on the fetus have, however, been areas of concern. The American Institute of Ultrasound in Medicine has published a list of conclusions regarding specifications and qualifications for its use. They conclude that, although the possibility exists that adverse effects may be identified in the future, "no confirmed biological effects on patients . . . caused by exposure at intensities typical of present diagnostic ultrasound instruments have ever been reported" (Lizzi 1988). In the same publication it was noted that "other epidemiologic studies have shown no causal association of diagnostic ultrasound with any of the adverse fetal outcomes studied." Lizzi (1988) adds that the acoustic exposure levels in these studies may not be representative of the full range of fetal exposures characteristic of current practices, a consideration that suggests a need for caution if exposure levels were to increase in the future.

Postnatal Monitoring for Hearing Loss

To date, studies carried out to identify postnatal hearing loss in children whose mothers were repeatedly exposed to high levels of noise during the child's prenatal development, have been inconclusive because of the inadequacy of sample sizes, control groups, and maternal noise exposure data (Gerhardt 1990; Lalande, Hétu, and Lambert 1986). Similar problems affect studies of premature babies (Henry 1983; Mitchell 1984; Minoli and Moro 1985) who have spent long periods following their early births in NICUs. Intensity levels of NICU incubator ambient noise can reach from 60 to 80 dB, and may be raised an additional 5 to 25 dB by other life support equipment and activities in the surrounding area, with up to 140 dB impulse levels occurring from time to time (Gerhardt 1990; Ryals 1990). Difficulties abound in assessing hazardous levels for both prenatal and postnatal environmental noise exposure and their possible synergistic effects in conjunction with ototoxic medications. The determination of appropriate time limits for such exposures is another problem to be solved. These areas have not been widely studied and represent a fertile field for controlled investigation.

Other Fetal Effects

The literature contains a few references to other possible ill effects on the fetus from exposure to noise. Studying a Japanese population, Nakamura (1977) has described reduced birthweight that may have been caused by exposure to noise; Schell (1981) has reported that noise may reduce prenatal growth in the human fetus; and Jones (1983) has suggested that a wide range of non-auditory effects involving fetal mal-development may result from noise exposure. On the other hand, Edmonds and his colleagues (1979), in their well-controlled investigation of the effects of airport noise on birth defects, found no significantly increased risk for residents of high-noise areas, stating that "it is our opinion that noise or other factors associated with residence near airports are unlikely to be important environmental teratogens [i.e., agents capable of causing the development of a deformed fetus]." Here again there are great difficulties in controlling for the many variables involved in such studies. Current and future technological events in conjunction with rigorous research controls promise improved insights into the study of both auditory and non-auditory effects of fetal development. Our appreciation for the mechanisms of genetic heterogeneity is expanding, and new information in this exciting arena can have a considerable impact on our understanding of human fetal vulnerability to noise stimulation.

REFERENCES

CORWIN, J.T., and COTANCHE, D.A. 1988. Regeneration of sensory hair cells after acoustic trauma. *Science* 240:1772-1774.

EDMONDS, L.D., LAYDE, P.M., and ERICKSON, J.D. 1979. Airport noise and teratogenesis. *Archives of Environmental Health* 34:243-247.

GERHARDT, K.J. 1990. Prenatal and perinatal risks. In *National Institutes of Health (NIH) Consensus Development Conference on Noise and Hearing Loss, Program and Abstracts.*

GERHARDT, K.J., ABRAMS, R.M., KOVAZ, B.M., GOMEZ, K.J., and CONLON, M. 1988. Intrauterine noise levels in pregnant ewes produced by sound applied to the abdomen. *American Journal of Obstetrics and Gynecology* 159:228-232.

HENRY, K.R. 1983. Abnormal auditory development resulting from exposure to ototoxic chemicals, noise and auditory restriction. In *Development of the Auditory Vestibular System,* ed. R. Romand. New York: Academic Press.

JONES, F.N. 1983. Non-auditory effects of noise on fetal life. In *Proceedings of the Fourth International Congress on Noise as a Public Health Problem* (Turin), Vol. 1, ed. G. Rossi. Milan: Edizioni Techniche a cura del Centro Ricerche e Studi Amplifon.

KISILEVSKY, B., MUIR, D.W., and LOW, J. 1989. Human fetal responses to sound as a function of stimulus intensity. *Obstetrics and Gynecology* 73:971-975.

KUHLMAN, K.A., BURNS, K.A., DEPP, R., and SABBAGHA, R.E. 1988. Ultrasonic imaging of normal fetal response to external vibratory stimulation. *Obstetrics and Gynecology* 158:47-51.

LALANDE, N.M., HÉTU, R., and LAMBERT, J. 1988. Is occupational noise exposure during pregnancy a risk factor of damage to the auditory system of the fetus? *American Journal of Industrial Medicine* 10:427-435.

LAVIGNE-REBILLARD, M., and PUJOL, R. 1988. Hair-cell innervation in the fetal human cochlea. *Acta Otolaryngologica* 105:398-402.

LIZZI, F.L. 1988. Bioeffects considerations for the safety of diagnostic ultrasound. *Journal of Ultrasound* 7:1-38.

MINOLI, J., and MORO, G. 1985. Constraints of intensive care units and follow-up studies in prematures. *Acta Otolaryngologica* (Stockholm) (Supplement) 421:62-67.

MITCHELL, S.A. 1984. Noise pollution in the neonatal intensive care nursery. In *Early Identification of Hearing Loss in Infants,* ed. T. Makensy. *Seminars in Hearing* 5:17-25.

NAKAMURA, R. 1977. Gestation and noise. (Abstract) In *Congress Handbook, Seventh Asian Congress of Obstetrics and Gynecology,* eds. S. Toongsuwan and T. Suvonnakoto. Bangkok.

OHEL, G., HOROWITZ, E., LINDER, N., and SOHMER, H. 1987. Neonatal auditory acuity following in-utero vibratory acoustic stimulation. *American Journal of Obstetrics and Gynecology* 157:440-441.

QUERLEU, D., REYNARD, X., VERSYP, F., PARIS-DELRUE, L., and CRÉPIN, G. 1988. Fetal hearing. *European Journal of Obstetrics, Gynecology and Reproductive Biology* (Supplement) 28:191-212.

RAJAN, R., and JOHNSTONE, B.M. 1988. Binaural acoustic stimulation exercises protective effects at the cochlea that mimic the effects of electrical stimulation of an auditory efferent pathway. *Brain Research* 459(2):241-255.

RUBEL, E.W., and RYALS, B.M. 1983. Development of the place principle: acoustic trauma. *Science* 219:512-514.

RYALS, B.M. 1990. Critical periods and acoustic trauma. In *National Institutes of Health (NIH) Consensus Development Conference on Noise and Hearing Loss, Program and Abstracts.*

RYALS, B.M., and RUBEL, E.W. 1988. Hair-cell regeneration after acoustic trauma in adult coturnix quail. *Science* 240:1774-1776.

SAUNDERS, J.C., DEAR, S.P., and SCHNEIDER, M.E. 1985. The anatomical consequences of acoustic injury: a review and tutorial. *Journal of the Acoustical Society of America* 78(3):833-860.

SCHELL, L.M. 1981. Environmental noise and human prenatal growth. *American Journal of Physical Anthropology* 56:156-163.

SMITH, C.V., PHELAN, J.P., NGUYEN, H.N., JACOBS, N., and PAUL, R.H. 1988. Continuing experience with the fetal acoustic stimulation test. *Journal of Reproductive Medicine* (Supplement) 33:365-368.

VINCE, M.A., ARMITAGE, S.E., BALDWIN, B.A., TONER, J., and MOORE, B.C.J. 1982. The sound environment of the foetal sheep. *Behavior* 81:296-315.

Susceptibility to Noise-Induced Hearing Loss and Acoustic Trauma

ANITA T. PIKUS

Terminology

Human susceptibility to both noise-induced hearing loss and acoustic trauma varies widely, as does the degree of impairment imposed by each on the individual. While the exact differences in the physiological and biochemical events that occur in each are not fully understood, both processes can result in permanent hearing loss. In what follows, the use of the terms *noise-induced hearing loss* and *acoustic trauma* as referring to specific but different entities is consistent with the concept and language used in the statement drafted in January, 1990 at the Consensus Development Conference on Noise and Hearing Loss held at the National Institutes of Health (Noise and Hearing Loss 1990).

Endogenous Factors in Susceptibility to Noise

Individual Effects. Most of the early research in susceptibility has been centered on those variables that are easily quantifiable, such as different types of noise stimulation or exposure, ototoxic drug administration, and whole body vibration (Humes 1984). There have also been a few reports concerning the interactions of these noxious stimuli in human subjects (Fechter 1988). Other studies of individual susceptibility to hearing damage caused by noise have examined its effects on those with preexisting hearing losses of various types, or with other abnormal conditions in the middle or inner ear (Popelka 1990). Hearing damage arising from the use of hearing aids has also been explored, but the literature in this particular area is somewhat conflicting, and clearcut recommendations concerning individual prophylaxis cannot yet be made (Popelka 1990). Other approaches in studying individual variability in developing hearing loss from noise exposure have included analysis of the premise that specific eye color and skin pigmentation may act as predisposing factors (Cunningham and Norris 1982; Humes 1984, 1990). However, the trend in the current literature suggests that unequivocal evidence has not been documented to sustain this thesis.

Group Effects. Other researchers have suggested that differences in susceptibility to noise-induced hearing loss do exist among human racial groups, although these differences have not been successfully reproduced in animal models. Among racial groupings, it has been consistently found that blacks show less permanent threshold shift and noise-induced hearing loss than do whites (Jerger et al. 1986). Research on racial and sex differences in auditory sensitivity extends back to the early 1930s (Bunch and Raiford 1931). Research design, data analysis, and interpretation of results in these studies are often complicated by variables of the following sorts: occupational or recreational noise exposure, educational or socioeconomic status, preexisting health problems, and age. However, one study in particular was able to control for many of the contaminating study variables quite well, and data derived from this investigation do suggest that black males suffer less rapid or less severe noise-induced hearing loss than do white males receiving the same extent and rate of noise exposure (Jerger et al. 1986).

Age Effects. The effects of age on susceptibility to permanent hearing loss due to noise exposure are not well understood at either end of the human life span, where they appear to be most critical. Studies of susceptibility to noise early in the life span are based on animal subjects because of the difficulties encountered in human fetal hearing measurements. With respect to the developing fetus, Ryals (1990) has discussed the concept of a critical period of heightened susceptibility to cochlear damage from excessive noise exposure prior to the complete development of the auditory efferent system. Such periods have been found to exist in various animal species. They occur at specific but different times as a function of the differing rates of development for each species, but they always occur late in fetal life. The existence of such a period in human beings remains hypothetical. However, even the possibility of its existence in the late stages of fetal development should not be overlooked in the case of survivors of premature birth, in whom the heightened susceptibility could possibly extend into postnatal life, with its assured exposure to the potentially damaging noise levels commonly encountered in the neonatal intensive care unit (NICU).

In further studies of animal subjects, Henry (1982) reported on the influence of genotype and age on noise-induced auditory deficits using specific mouse genotypes. Age-related hearing loss was noted in all mice, in both the youngest subjects, which were the most severely affected, and again in those of increasing age, in which the maximum loss was noted at increasingly higher frequencies. The results of Mills' studies (1990) of Mongolian gerbils raised in quiet conditions demonstrated hearing loss in the last one-third of life, with audiometric configurations similar to those of 60- to 70-year-old human subjects. Hearing thresholds were significantly poorer for a similar group of Mongolian gerbils raised in noise levels of 85 dBA than for that raised in quiet. The hearing loss found in the "quiet" group was accounted for by changes in the peripheral ear, chiefly the degeneration of the stria vascularis, with some loss of both sensory and spiral ganglion cells. The greater loss in the noise-exposed animals was considered to be the sum of the hearing loss due to aging and that due to noise exposure, since subtraction of the quiet group's measures resulted, for the group exposed to noise, in an audiometric configuration that exhibited a correlation with the spectral features of the noise to which they had been exposed (Mills 1990).

Rosen and his colleagues (1962) have reported results of audiometric studies of aged human subjects in a primitive culture distantly removed from industrialized societies. In their virtually noise-free environment—the loudest noises they heard were their own singing and the crowing of roosters—the subjects exhibited normal hearing into their nineties. These findings, coupled with those of the studies of animal subjects, suggest that some age-related hearing loss in industrialized societies may be in large measure a result of noise damage, or of synergistic effects of noise with certain aging processes. The early and late effects of aging upon susceptibility to noise needs more well-controlled documentation before definitive statements may be made about them.

Gender effects. According to the latest National Institutes of Health (NIH) Consensus Development Conference on Noise and Hearing Loss held in 1990, gender differences in human hearing ability do not appear until after age 10, when hearing in females emerges as consistently better than that of males, a difference that persists into old age. Participants in the Consensus Conference concluded from available data that the differences are probably due to greater exposure of males to noise rather than to any inherent susceptibility on the part of males (Noise and Hearing Loss 1990). It is not clear at this time whether such an interpretation will be supported empirically.

Exogenous Factors and Susceptibility

Pharmacologic Therapies. Some of the susceptibility to hearing damage from noise exposure in individuals can be attributed to the use of ototoxic drugs and chemotherapeutic agents (Dayal et al. 1971; Gannon, Tso, and Chung 1979). According to the most recent research, aminoglycosides may, in fact, increase susceptibility to noise damage (Salvi 1990). This has not been clearly demonstrated in human subjects, but research with animals has unequivocally implicated aminoglycosides. And cisplatin, a heavy-metal compound used in cancer treatment and known to be ototoxic, can significantly increase hearing loss in combination with noise exposure. Hence, new precautions are being observed with respect to noise-induced hearing loss for patients who are being treated with certain antibiotic or other pharmacologic therapies (Gratton et al. 1988). Salicylates (including aspirin) are the most widely used drug in the industrialized societies, and high doses can cause temporary tinnitus and hearing loss (Salvi 1990). Studies with animal subjects have shown that combinations of salicylates and noise exposure increase the severity of temporary threshold shift only slightly over that produced by either condition alone (Woodford, Henderson, and Hamernik 1978; McFadden and Plattsmier 1983). However, none of the studies has shown that the risk of permanent threshold shift from noise exposure is increased by salicylates (Salvi 1990).

Chemical agents. On the strength of four case studies, Barregård and Axelsson (1984) have suggested that painters who are exposed to organic solvents (e.g., carbon disulfide and *n*-butanol) while working in noisy places such as shipyards, may sustain unusually severe noise-induced hearing losses due to the combination of the two conditions. Morata (1989) found similar results for Brazilian workers in a rayon factory who were simultaneously exposed to carbon disulfide and noise. When their data were adjusted for subjects' ages, Ödkvist and his coworkers (1987) did not find that hearing sensitivity was affected in subjects exposed to industrial solvents and to noise. Rather, the major abnormal auditory effects were retrocochlear in origin, and pathological responses to vestibular and oculomotor tests were found. Other chemical agents such as carbon monoxide, toluene, trichloroethylene, xylene, and styrene have also been implicated as synergists with noise (Fechter 1988; Morata 1989; Salvi 1990). Determinations of exposure threshold limits, if any, for these agents have not been based on considerations of the possible effects of simultaneous exposures, and extensive study of these unique interactions is needed (Morata 1989).

Vibration and Susceptibility. High-level vibration frequently accompanies noise in the workplace, but does not appear to be a cause of permanent hearing loss by itself (Hamernik et al. 1981). However, more noise-induced hearing loss has been reported in certain workers exposed to both noise and vibration than to noise alone (Pykko et al. 1981; Iki et al. 1985).

Conclusion

Unfortunately, in humans, the genetic bases underlying individual differences in susceptibility to noise damage have not yet been studied adequately. Investigations into the relation of these genetic bases to age, ototoxic agents, gender differences, biochemical variables, pigmentary differences, and anomalies in certain other organ systems have been almost completely neglected. Currently, there is no method by which to sort out the relative contribution of any of these factors in order to identify persons who have sustained or are at risk for the most severe or earliest forms of hearing loss due to noise exposure. The National Institute on Deafness and Other Communication Disorders concluded in its draft statement following its Consensus Development Conference on Noise and Hearing Loss that "research on the molecular mechanisms of the development of noise-induced hearing loss, the genetic basis of susceptibility to noise-induced hearing loss, and the effectiveness of educational and hearing conservation programs is needed." They also concluded that the state of the art of our scientific knowledge makes risk assessment for any one individual impossible (Noise and Hearing Loss 1990).

The future of research in genetic susceptibility to noise-induced hearing loss promises to be greatly facilitated by newly available clinical techniques measuring cochlear emissions. (It is known that cochlear emissions are highly stable over time in normal subjects and, more important, each individual is thought to produce a unique and repeatable response.) This technology offers a whole range of study possibilities for clinical correlations in the investigation of the genetics of noise-induced hearing loss.

REFERENCES

Barregård, L., and Axelsson, A. 1984. Is there an ototraumatic interaction between noise and solvents? *Scandinavian Audiology* 13:151-155.

Bunch, C.C., and Raiford, T.S. 1931. Race and sex variations in auditory acuity. *Archives of Otolaryngology* 13:423-434.

Cunningham, D.R., and Norris, M.L. 1982. Eye color and noise-induced hearing loss: a population study. *Ear and Hearing* 3:211-214.

Dayal, V.S., Kokshanian, A., and Mitchell, D.P. 1971. Combined effects of noise and Kanamycin. *Annals of Otology, Rhinology and Laryngology* 80:1-6.

Fechter, L. 1988. Prediction of synergistic effects of environmental agents on auditory function. In *Recent Advances in Research on the Combined Effects of Environmental Factors,* ed. O. Manninen. Tampere, Finland: Py-Paino Oy Printing House.

Gannon, R.P., Tso, S.S., and Chung, D.Y. 1979. Interaction of Kanamycin and noise exposure. *Journal of Laryngology and Otology* 93(4):341-347.

Gratton, M.A., Salvi, R.J., Kamen, B.A., Henderson, D., and Poon, M. 1988. Combined effects of noise and cisplatin: Effect on hearing. In *Recent Advances in Research on the Combined Effects of Environmental Factors,* ed. O. Manninen. Tampere, Finland: Py-Paino Oy Printing House.

Hamernik, R.P., Henderson, D., Coling, D., and Salvi, R. 1981. Influence of vibration on asymptotic threshold shift produced by impulse noise. *Audiology* 20:259-269.

Henry, K.R. 1982. Influence of genotype and age on noise-induced auditory losses. *Behavior Genetics* 12(6):563-572.

Henry, K.R. 1982. Influence of genotype and age on noise-induced auditory losses. *Behavior Genetics* 12(6):563-572.
Humes, L.E. 1984. Noise-induced hearing loss as influenced by other agents and by some physical characteristics of the individual. *Journal of the Acoustical Society of America* 76:1318-1329.
Humes, L.E. 1990. Individual susceptibility—nonauditory factors. In *National Institutes of Health (NIH) Consensus Development Conference on Noise and Hearing Loss, Program and Abstracts.*
Iki, M., Kurumatani, N., Hirata, K., and Moriyama, T. 1985. An association between Raynaud's phenomenon and hearing loss in forestry workers. *American Industrial Hygiene Association Journal* 46(9):509-513.
Jerger, J., Jerger, S., Pepe, P., and Miller, R. 1986. Race differences in susceptibility to noise-induced hearing loss. *The American Journal of Otology* 7(6):425-429.
McFadden, D., and Plattsmier, H.S. 1983. Aspirin can potentiate the temporary hearing loss induced by intense sounds. *Hearing Research* 9:295-316.
Mills, J.H. 1990. Noise and the aging process. In *National Institutes of Health (NIH) Consensus Development Conference on Noise and Hearing Loss, Program and Abstracts.*
Morata, T.C. 1989. Study of the effects of simultaneous exposure to noise and carbon disulfide on workers' hearing. *Scandinavian Audiology* 18:53-58.
Noise and Hearing Loss, NIH Consensus Development Conference Statement. 1990. 8(1).
Ödkvist, L.M., Arlinger, S.D., Edling, C., Larsby, B., and Bergholtz, L.M. 1987. Audiological and vestibulo-oculomotor findings in workers exposed to solvents and jet fuel. *Scandinavian Audiology* 16:75-81.
Popelka, G.R. 1990. The effects of certain auditory factors on individual susceptibility to noise. In *National Institutes of Health (NIH) Consensus Development Conference on Noise and Hearing Loss, Program and Abstracts.*
Pykko, I., Starch, J., Farkkila, M., Hoikkala, M., Korhonen, O., and Nurminen, M. 1981. Hand-arm vibration in the aetiology of hearing loss in lumberjacks. *British Journal of Industrial Medicine* 38:281-289.
Rosen, S., Bergman, M., Plester, D., El-Mofty, A., and Satti, M. 1962. Presbycusis study of a relatively noise-free population in the Sudan. *Annals of Otology, Rhinology and Laryngology* 71:727-743.
Ryals, B.M. 1990. Critical periods and acoustic trauma. In *National Institutes of Health (NIH) Consensus Development Conference on Noise and Hearing Loss, Program and Abstracts.*
Salvi, R. 1990. Interaction between noise and other agents. In *National Institutes of Health (NIH) Consensus Development Conference on Noise and Hearing Loss, Program and Abstracts.*
Woodford, C.M., Henderson, D., and Hamernik, R.P. 1978. Effects of combinations of sodium salicylate and noise on the auditory threshold. *Annals of Otology, Rhinology and Laryngology* 87:117-127.

The Effects of Noise on the Ear and Hearing

BARBARA ASHKINAZE AND MARC B. KRAMER

It has long been a recognized fact that long-term exposure to noise can cause sensorineural hearing loss. Modern techniques of measurement and analysis have progressed to the point where quantitative dose-response relationships have been derived between exposure level and duration of noise and the expected degree of damage (DeJoy 1984).

The onset of loss of auditory sensitivity due to noise exposure may be related to a specific incident, but it is more often a slow process that takes place cumulatively over a period of years. Not all exposure to hazardous noise levels occurs at the workplace and not all individuals are equally susceptible to the deleterious effects of noise.

Traditionally, studies of noise exposure have included after-the-fact analyses of industrial noise and measurement of the resultant hearing loss. A principal complication of such studies lies in the uncertainty of reconstructing an individual's unique history of noise exposure which also includes any exposure to hazardous noise outside of the work situation in everyday life (sociacusis). A second complication arises from the fact that symptoms similar to those of noise-induced hearing loss can also be caused by such factors as the aging process (presbyacusis) and ototoxic drugs (nosoacusis).

Laboratory studies of the human response to noise provide a better control over noise exposure variables; however they are usually restricted to exposures of short duration (several minutes to a few hours) and for ethical reasons are limited to exposures that produce relatively low levels of temporary threshold shift. We thus face a methodological problem: we are restricted to employing experiments that induce a minor hearing loss over a short period of time in the effort to understand the sort of process that may underlie a serious hearing impairment that has developed over many years (Henderson, Hamernik, and Hynson 1979).

Loss of Auditory Sensitivity

Temporary Threshold Shift (TTS). The primary effect of noise exposure is hearing loss. Studies have shown that when a normal ear

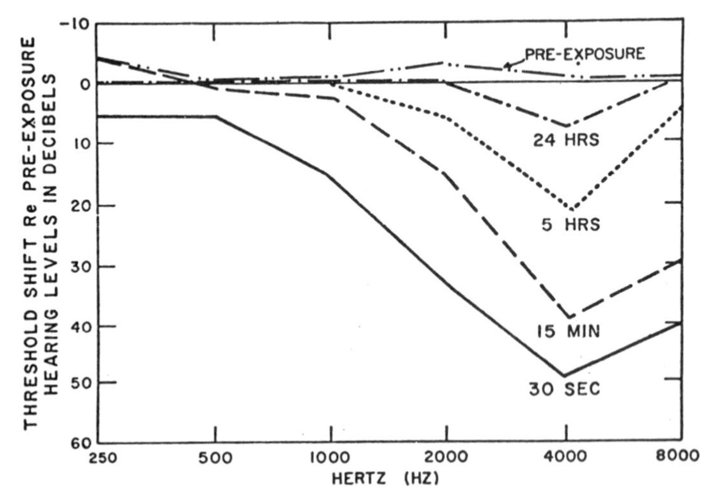

FIGURE 1. Single subject hearing levels measured at various times after a 2-hour exposure to a broad-band noise at 103 dBA as compared with pre-exposure determinations. (From *Industrial Noise: A Guide to Its Evaluation and Control* 1967. Department of Health, Education, and Welfare, Public Health Service, Publication Number 1572.)

is exposed to continuous noise levels above 80 dBA for long periods of time, a temporary decrease in hearing sensitivity occurs. This temporary hearing loss is called *temporary threshold shift* (TTS), *noise-induced temporary threshold shift* (NITTS), or *auditory fatigue* (Mitchell 1984). When the maximum energy of a sound is concentrated in its low-frequency components, less TTS will result than when the maximum energy is concentrated in its high-frequency components (American Medical Association 1973). Stephenson and his colleagues exposed subjects to continuous noise for 24 hours at 65, 70, 75, 80, and 85 dBA. TTS growth and subsequent recovery were measured at specific intervals throughout each exposure. Results demonstrated that at the frequency to which greatest sensitivity was exhibited (4 kHz), a TTS could be predicted for a signal intensity exceeding 80 dBA (Stephenson, Nixon, and Johnson 1980). Typically, TTS occurs within the first hour or two following exposure. The rate of recovery is dependent upon the magnitude of the initial loss, and generally speaking, the greater the initial loss is, the faster the recovery will be. However, when TTS reaches 40 dB or more, recovery may be slowed, with TTS requiring days or even weeks to disappear. This 40-50 dB range of TTS may represent a critical TTS that should not be exceeded if danger of permanent damage to hearing is to be avoided. FIGURE 1 shows a single subject's pre-exposure normal hearing levels as compared to those measured at various time intervals following exposure for two hours to a broad-band noise of 103 dBA

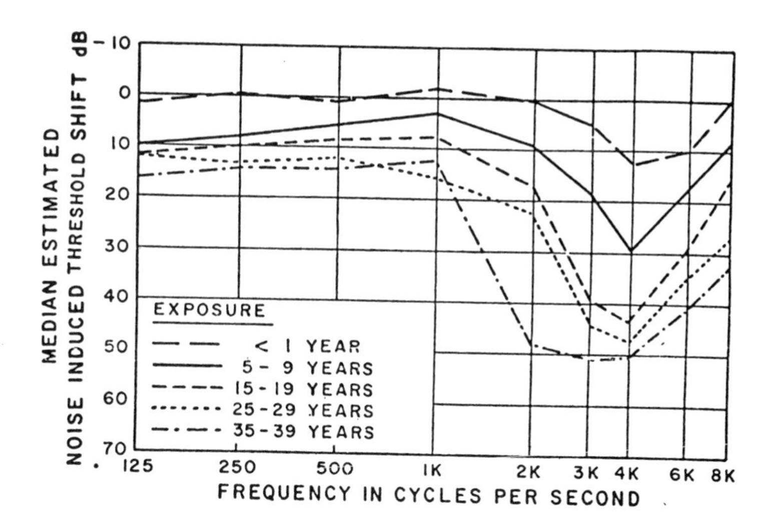

FIGURE 2. Median noise-induced permanent threshold shifts (NIPTS) in hearing levels as a function of exposure years to jute weaving noise. (From *Industrial Noise: A Guide to Its Evaluation and Control* 1967. Department of Health, Education, and Welfare, Public Health Service, Publication Number 1572.)

intensity. Note that by far the greatest amount of recovery occurred during the first 15 minutes following the exposure. Recovery was still not complete 24 hours later.

Variability in TTS has been reported both between individuals as well as within the same individual across time, even with identical noise exposure (Humes 1980). One possible cause contributing to the variability in TTS is individual differences in subjective experience of the fatiguing stimulus (Swanson et al. 1987). Lindgren and Axelsson (1983) compared the effects of noise and music on hearing and concluded that high sound levels experienced as distressing or noxious cause more TTS than high sound levels that the listener finds enjoyable. Other studies have reported that the magnitude of TTS is greater in situations that promote subjective distress. Hormann, Mainka, and Gummlich (1970) observed greater TTS in subjects who listened to a noise that served as a punishment for a task error than in control subjects. The full relationship between temporary threshold shift and permanent threshold shift, as well as the effect of listener attitude toward the stimulus, are still not clear.

Permanent Threshold Shift. The cumulative effects of long-term exposure to high noise levels lead to permanent hearing loss referred to as *noise-induced permanent threshold shift* (NIPTS), or simply as *permanent threshold shift* (PTS). FIGURE 2 shows the median NIPTS in hearing levels for groups of workers exposed to jute weaving noise over various periods of time, ranging from less than one year up through 39 years. Note the degree of loss at 2 kHz in the group exposed for periods of 35 to 39 years. This will result in their having difficulty understanding the speech of others, even in quiet conditions. The onset of NIPTS is characteristically gradual, but can also result from acoustic trauma such as a single exposure to the high-intensity noise of an explosion. NIPTS is progressive with continued exposure and is often accompanied by a ringing tinnitus, a sensation of fullness in the ears, and distortion of perceived speech. Pure-tone audiometry usually reveals a bilaterally symmetrical notched hearing threshold contour with the greatest loss in the 3-6 kHz region. With continued noise exposure, as the loss at the higher frequencies increases, the lower frequencies also become involved. Speech discrimination ability varies depending on the frequencies affected and the magnitude of the loss. When a hearing loss is confined to the frequencies above 3 kHz, maximum speech discrimination ability may be unimpaired in quiet conditions. However, as the frequencies of 3 kHz and below become involved, speech discrimination ability decreases as the degree of pure-tone loss increases (Jerger and Jerger 1981).

The amount of NIPTS in a given individual is dependent upon that person's unique susceptibility as well as the nature of the noise stress, including its overall sound pressure level, its frequency composition, and its duration.

Biological Bases of Noise-Induced Hearing Loss

Research conducted over the past thirty years has identified the degeneration of sensory hair cells in the cochlea as the anatomical damage that results from exposure to excessive noise. The hair cells convert the mechanical energy of sound vibrations into neuroelectrical signals. The actual physiological mechanisms that result in hair cell death are as yet unknown, but the process of cell degeneration has been well documented (Spoendlin 1972).

A great deal of attention has been paid to the effect of noise on cochlear blood flow. Current thinking suggests that the inherently small blood supply to the organ of Corti in the cochlea limits the cochlea's capacity for rapid exchange of cell nutrients and waste products. Because of this limitation, even relatively short periods of sound overstimulation that induce vasoconstriction may cause cellular depletion. Results of studies have shown a variety of effects, from no change at all to either an increase or a decrease in blood flow (Axelsson and Vertes 1982). It is known that one of the earliest detectable signs of this pathologic process is the swelling of the hair cells. This change usually results in damage to the outer rows of hair cells with a gradient of damage to the inner row. Paradoxically, however, studies of acoustic trauma have quite often found hearing loss, as assessed by physiological or behavioral methods, that is not accompanied by observable hair cell loss.

Recent studies of acoustic injury to the auditory nervous systems of animals suggest that noise-induced hearing loss (NIHL) may also receive important contributions from retrocochlear lesions in the central nervous system (CNS). Studies suggesting CNS functional loss that is independent of the cochlea remain inconclusive because of the difficulty of excluding influences of the peripheral ear. Hence, data from direct anatomical studies of the animal CNS after exposure of the subject to noise remain the only convincing evidence implicating the central nervous system in NIHL.

At various locations along the central auditory pathways, cellular changes, such as reduced cell size and irregularities in the nuclei that lead to cell degeneration, have been found in various species, suggesting that overstimulation results in changes in auditory neuronal circuitry that are not found in unexposed animals (Saunders, Dear, and Schneider 1985). Further study of these changes is necessary, as is the determination of their role in PTS.

Parameters of Hazardous Noise Exposure

Three major factors that contribute to the harmfulness of excessive noise exposure have emerged from available laboratory data and epidemiological studies of industrial populations. They are the overall intensity of the noise, its frequency composition, and the duration of the exposure.

Intensity. Noise-induced hearing loss increases as the intensity of the noise increases. The risk of hearing loss increases as the noise exposure level exceeds 90 dBA. In a study conducted by Ward and his colleagues, groups of chinchillas were subjected to a series of noise exposures of approximately equal energy ranging from 22 minutes at 120 dBSPL to 150 days at 82 dBSPL. For all exposures involving levels of 112 dB or less, the same average permanent hearing loss (15-20 dB) and degree of outer hair cell destruction (8-10%) resulted. The 22-minute exposure at 120 dB produced a 60 dB hearing loss and massive hair cell destruction (70-80%), indicating that some threshold critical with respect to acoustic trauma had been exceeded. (Ward, Santi, and Duvall 1981).

Frequency. With exposure to intense noise, the portion of the ear that is damaged, and hence the frequencies at which hearing loss occurs, is determined to a large extent by the frequency composition of the noise (Moody and Stebbings 1986). When the spectrum of noise is broad-band and relatively equal in energy over the entire bandwidth, the maximum shift occurs in the frequency range between 3,000 and 6,000 Hz. When the noise is contained in a narrower bandwidth, the frequencies affected depend on the range of frequencies in that band. Generally, the higher the exposure frequency, the greater the amount of threshold shift. Mills and his associates exposed groups of human subjects to octave band noise centered at 63, 125, or 250 Hz for periods of eight or 24 hours. For the 250 Hz condition, TTS increased about 1.5 dB per decibel increase in noise level, whereas for the 63 and 125 Hz conditions, TTS increased less than 1 dB per decibel increase (Mills et al. 1983). This observed relationship is reflected in most damage risk criteria, which permit exposure to greater levels of low-frequency noise than would be acceptable in the mid-frequency range of 600 to 2,400 Hz.

Duration. The duration of exposure to a noise is of equal importance to its intensity and frequency components, and the interaction of all three parameters must be considered in determining the effects of a given noise exposure on NIPTS.

Most noise encountered in the industrial environment is fluctuating; that is, the noise is continuous but the levels rise and fall by more than 5 dB during a given exposure period. Glorig divides the temporal parameters of noise into two dichotomous pairs: steady versus impulsive, and continuous versus intermittent. Both steady and impulsive noise can thus be classified as either continuous or intermittent. Steady noise that is continuous differs in terms of its effects on the ear when it is made intermittent (i. e., broken down into periods of "on" and "off" times) (Glorig 1980). A single impulse is usually heard as a discrete event, either occurring in otherwise quiet conditions, or superimposed on a background of steady-state continuous noise (Melnick 1985).

Early studies of occupational hearing loss suggested that a typical hearing-loss pattern would appear first in the 3-6 kHz range, with subsequent spread to the lower frequencies. However, Sataloff and his colleagues (1984) have subsequently reported that intermittent exposure to intense noise results in a very severe loss in the higher frequencies, with relatively little or no hearing loss in the lower frequencies, even after many years of exposure. This effect differs substantially from the effects of continuous exposure to noise of the same intensity. It remains to be determined whether this pattern of hearing loss results from intermittent exposure to all types of noise or only to the type investigated in this study. There has also been speculation that recovery during the off time is a contributing factor. Mäntyslo and Vuori investigated the effects of impulse noise and continuous steady-state noise on hearing. They concluded that as the duration of exposure to impulse noise increases, the region of affected frequencies becomes wider, with increased threshold shifts in both ears. Impulse noise seemed to produce permanent threshold shifts at 4 and 6 kHz after a shorter exposure duration than continuous steady-state noise (Mantysälo and Vuori 1984).

The criteria for hearing damage induced by steady-state noise are well established, but it is uncertain whether they are also valid for exposure to impulse noise. Stevin (1972) advocates extending to impulse noise the same noise dose concept that is used for continuous noise.

Assessment of Risk

The relationships that exist between exposures to sounds of differing intensities and durations and the risk of sustaining consquential hearing impairment have been called damage risk criteria (DRC). Melnick (1982) summarized the concepts from which such criteria have been derived, and described how they have traditionally been applied. The initial goal of such formulations was to establish a measurable noise level that could not be exceeded if assurance against loss of auditory sensitivity was to be avoided. After establishment of this limit (explicitly defined in terms of a specific hearing sensitivity impairment), further evaluation of the relationship between the degree of exposure (dose) and the magnitude of noise-induced hearing loss (reponse) was undertaken (Hodge and Price 1978).

The development of a damage risk document is a complex task, for which Melnick (1982) enumerated five important considerations.

1. How much hearing does one wish to preserve?
2. How does one measure hearing?
3. What is an acceptable degree of risk?
4. What is the relationship of the noise exposure to hearing loss?
5. How does one measure the noise or represent the noise environment?

Each of these factors must then be further assessed against such concerns as the purpose of the DRC's specific application and the practical restrictions imposed by fiscal limitations. Effects upon hearing of unrelated noise exposures, of the aging process, of systemic disease, and of the use of ototoxic medication or other substances capable of causing PTS must furthermore be taken into account. Finally, consideration must be given to the possibility of synergistic effects arising from the combination of noise with other forms of environmental hazard, for example, noxious fumes.

In general, the definition of each of these five basic factors is dependent upon that of each of the others. For example, the first,

TABLE I. Hearing Impairment Threshold (Low Fence) Values

Organization	Frequencies Averaged (kHz)	Low Fence (dB)
NIOSH (1972)	1.0, 2.0, 3.0	25
AAO (1979)	0.5, 1.0, 2.0, 3.0	25
AMA (1979)	0.5, 1.0, 2.0, 3.0	25
ASHA (1981)	1.0, 2.0, 3.0, 4.0	25

"How much hearing does one wish to preserve?" is immediately dependent upon the way in which the parameters of normal and impaired hearing are described and the methods by which hearing is measured. Pure-tone sensitivity as measured by threshold assessment has certainly achieved the widest acceptance, although there is still no universal agreement as to which measures provide the best descriptors of our ability to understand speech and to communicate (Melnick 1982). Definitions of hearing impairment based on such data have been offered by various professional groups, including the National Institutes of Occupational Safety and Health (NIOSH) (1972), the American Academy of Otolaryngology (AAO) (1979), the American Medical Association (AMA) (1979), the American Speech-Language-Hearing Association (ASHA) (1981), and others. In each case, a simple numerical average of the pure-tone threshold values obtained at specified frequencies has been chosen to represent the relative difficulty imposed by the impairment upon spoken communication (TABLE I). A specific threshold average below which hearing sensitivity is considered to be unimpaired, the so-called "low fence," has furnished the benchmark against which significant hearing loss may be judged for purposes of establishing DRC. A discussion of the practicality of this approach is beyond the scope of this publication, but consideration of the philosophy underlying this technique is strongly recommended.

Assessment of the noise levels associated with a given noise-induced hearing loss is accomplished by comparing the pure-tone auditory sensitivity of groups of individuals who are believed to have histories free of excessive noise exposures to that of groups who have been repeatedly exposed to excessive levels of noise of known acoustic characteristics.

As mentioned above, the cause-and-effect relationship that exists between continuous noise levels of given intensities and durations and the degree of loss of sensitivity is quite well-established. Unfortunately, much exposure to excessive noise levels involves not only continuous noise or intermittently continuous noise, but also impulse as well as impact noise, for which the causal relationships remain quite unclear. The advent of increasingly sophisticated audiodosimeters that have the ability to take into account brief occurrences of high-intensity exposure, as well as the ability to integrate all contributory data into a single equivalent time-weighted eight-hour exposure, will undoubtedly add to our predictive abilities.

Even with our ever-increasing knowledge regarding the relationships that exist between exposure to excessive noise levels and resultant NIPTS, Melnick (1982) aptly reminds us that damage risk criteria are statistical concepts, and are subject to social, political, and economic constraints in their development. And since statistics refer to populations, they should not be applied to individuals.

Readers particularly interested in DRC may consult the International Standards Organization's Standard 1999: Acoustic Assessment of Occupational Noise Exposure for Hearing Conservation Purposes (1980); the National Institute for Occupational Safety and Health (NIOSH) Criteria for a Recommended Standard ... Occupational Exposure to Noise (1972); and the U.S. Environmental Protection Agency summary documents, Public Health and Welfare Criteria for Noise (EPA 1973), and Information on Levels of Environmental Noise Requisite to Protect Public Health and Welfare With an Adequate Margin of Safety (EPA 1974). Extensive discussion of these and other documents relating to DRC can be found in the records of the hearings held on the topic of standards to be promulgated under the Noise Exposure amendment to the Occupational Safety and Health Act [29 USC 1910.95 et seq.] (Department of Labor, OSHA 1981a, 1981b, 1983).

Sources of Noise-Induced Hearing Impairment

While occupational noise exposure is certainly the primary cause of NIPTS, not all hazardous noise is a feature of the workplace; it may also emanate from recreational or domestic activities. Excessively amplified rock music, snowmobiles, electronic arcade games, toy guns, and other noise sources, have been reported as posing a potential detriment to hearing (Plakke 1983). Even toys given to infants and very young children are capable of exposing them to sound at levels well in excess of 100 dBA, if the sound source is held directly to the ear, as is a common practice among children. Although noise standards have been set for industry, there are, to date, few noise standards for recreation, and none for toys.

Studies have confirmed that overall sound pressure levels of loud music, either at concerts or from personal stereos, frequently exceed current hearing damage risk criteria. These noise levels have produced large amounts of threshold shifts in both musicians and listeners. At some performances, sound levels near the band have been recorded at 115 dBA (EPA 1973). Personal radio/cassette headphones can produce peak outputs of 120 dBA at the ear (Axelsson 1990). In a study conducted by Lee and his colleagues (1985), 16 volunteers listened to headphone sets for three hours at their usual maximal level. Transient threshold shifts of 10 dB were observed in six volunteers and 30 dB in one volunteer. Catalano and Levin (1985) studied 190 public college students in New York City via a self-administered questionnaire regarding volume setting and weekly exposure in hours to portable radio/cassette players. Based on OSHA criteria for the permissible noise dose in the workplace, the noise exposure for 31.4% of the students equalled or exceeded the maximum allowable dose, and 50% of the group considered to be at risk exceeded the risk criteria by more than 100%.

Noise-induced temporary threshold shift has been documented in rock-and-roll devotees, but the incidence of hearing loss is moderate. A principal contributory factor in this apparent discrepancy is the characteristic intermittency of the music, with on-times of approximately three to five minutes alternating with off-times of approximately one minute. Several musicians such as Rod Stewart, Peter Townshend, Ted Nugent, and Alex Van Halen have publicly acknowledged that they suffer from tinnitus and hearing loss (Axelsson 1990). Sound levels in excess of 100 dBA are routinely encountered by attendees at noisy discotheques (Clark 1990), and rock concert levels can reach 115 dBA. Speakers now in use that are electrically equipped with up to 500,000 watts of power can emit sound levels capable of producing immediate acoustic trauma resulting in permanent hearing loss (Axelsson 1990).

A few common household devices produce levels that should cause concern, notably gasoline-powered leaf blowers, with levels measured up to 112 dBA, and chain saws, whose output can reach up to 116 dBA. Hearing protection should be worn when operating these devices. Cordless telephone ringers can produce sound pressure levels up to 140 dB and there are reports of individuals sustaining permanent hearing loss after incidents of a single exposure (Clark 1990).

The most serious threat of hearing impairment posed by leisure-time activities is associated with recreational hunting and target shooting. Peak sound levels from rifles and shotguns have ranged from 132 to 170 dBA. Clinical reports concerning hearing loss following exposure to shooting noise can be found in the literature dating back to the nineteenth century. Studies of the relationship between hearing loss and avocational shooting indicate, for the ear ipsilateral to the firearm, an asymmetrical auditory threshold of from 15 to 30 dB at 3, 4, and 6 kHz (Clark 1990).

The wearing of personal hearing protective devices (HPDs) in their various forms is a widely used preventive measure. Ear plugs of various types and effectiveness are readily available. Care should be exercised in their use, and professional guidance should be sought concerning their proper insertion and the possible consequences of their continued use.

There are ear inserts with special characteristics that provide varying degrees of hearing protection. For example, there is an HPD containing a diaphragm that protects the wearer by closing momentarily when it is struck by the shock wave of a gunshot or other impulse sound, but this sort does not always provide protection from high-level steady-state sounds.

There are also types of HPDs made of special elements that provide protection for musicians and others who need to hear with precision while exposed to sound levels reaching 120 dBSPL. These devices are not intended to provide maximum hearing protection. For that purpose, ear muffs of varying attenuation characteristics may be worn alone or in conjunction with ear plugs.

For an excellent summary of the state of the art in HPDs, and for an outline of research priorities in this connection, the reader is referred to the chapter by Elliot Berger in the Program and Abstracts of the 1990 NIH Consensus Development Conference on Noise and Hearing Loss (Berger 1990).

REFERENCES

American Academy of Otolaryngology (AAO), Committee on Hearing and Equilibrium; and the American Council of Otolaryngology Committee on the Medical Aspects of Noise 1979. Guide to the evaluation of hearing handicap. *Journal of the American Medical Association* 241:2055-2059.

American Medical Association (AMA) 1973. The Physicians Guide to Noise Pollution. Chicago: AMA.

American Speech-Language-Hearing Association (ASHA) 1981. Report of the task force on the definition of hearing handicap. *Asha* 23:293-297.

Axelsson, A. 1990. Noise exposure in adolescents and young adults. In *National Institutes of Health (NIH) Consensus Development Conference on Noise and Hearing Loss, Program and Abstracts.*

Axelsson, A., and Vertes, D. 1982. Histological findings in cochlear vessels after noise. In *New Perspectives on Noise-Induced Hearing Loss,* eds. R.P. Hamernik, D. Henderson, and R. Salvi. New York: Review Press.

Berger, E.H. 1990. Hearing protection—the state of the art (circa 1990) and research priorities for the coming decade. In *National Institutes of Health (NIH) Consensus Development Conference on Noise and Hearing Loss, Program and Abstracts.*

Catalano, P.J., and Levin, S.M. 1985. Noise-induced hearing loss and portable radios with headphones. *Journal of Pediatric Otorhinolaryngology* 9:59-67.

Clark, W. 1990. Noise exposure and hearing loss from leisure activities. In *National Institutes of Health (NIH) Consensus Development Conference on Noise and Hearing Loss, Program and Abstracts.*

DeJoy, D.M. 1984. The non-auditory effects of noise: Review and perspectives for research. *Journal of Auditory Research* 24:123-150.

Department of Labor, Occupational Safety and Health Administration (OSHA) 1981a. Occupational Noise Exposure; Hearing Conservation Amendment. *Federal Register* 46:4078-4179.
Department of Labor, Occupational Safety and Health Administration (OSHA) 1981b. Occupational Noise Exposure; Hearing Conservation Amendment, Rule and Proposed Rule. *Federal Register* 46:42622-42639.
Department of Labor, Occupational Safety and Health Administration (OSHA) 1983. Occupational Noise Exposure; Hearing Conservation Amendment, Final Rule. *Federal Register* 48: 9738-9783.
EPA (U.S. Environmental Protection Agency) 1973. Public health and welfare criteria for noise. EPA 550/9-73-002, Washington, D.C.
EPA (U.S. Environmental Protection Agency) 1974. Information on the levels of environmental noise requisite to protect public health and welfare with an adequate margin of safety. EPA 550/9-74-004, Washington, D.C.
Gloric, A. 1980. Noise, past, present, and future. *Ear and Hearing* 1(1):4-18.
Henderson, D., Hamernik, R.P., and Hynson, K. 1979. Hearing levels from simulated workweek exposure to impulse noise. *Journal of the Acoustical Society of America* 65(5): 1231-1237.
Hodge, D.C., and Price, G.R. 1978. Hearing damage risk criteria. In *Noise and Audiology*, ed. D.M. Lipscomb. Baltimore: University Park Press.
Hormann, H., Mainka, G., and Gummlich, H. 1970. Psychische und physische Reaktionen auf Geraüsch verschiedener subjektiver Wertigkeit. *Psychologische Forschung* 33:289-309.
Humes, L.E. 1980. Susceptibility to TTS: A review of recent developments. In *Proceedings of the Third International Congress on Noise as a Public Health Problem* (Freiburg), *ASHA Reports 10*, eds. J. Tobias, G. Jansen, and W.D. Ward. Rockville, Maryland: American Speech-Language-Hearing Association.
Industrial Noise: A Guide to its Evaluation and Control 1967. Department of Health, Education, and Welfare, Public Health Service, Publication Number 1572.
International Standards Organization (ISO) 1980. Assessment of occupational noise exposure with respect to hearing impairment. Draft Proposal 1999/1, Technical Committee 43, Subcommittee 1, Work Group 19.
Jerger, S., and Jerger, J. 1981. Noise-induced hearing loss. In *Auditory Disorders: A Manual for Clinical Evaluation.* Boston: Little, Brown and Company.
Lee, P.C., Senders, C.W., Gantz, B.T., and Otto, S.R. 1985. Transient sensorineural hearing loss after overuse of portable headphone cassette radios. *Otolaryngology, Head and Neck Surgery* 93:62-65.
Lindgren, F., and Axelsson, A. 1983. Temporary threshold shift after exposure to noise and music of equal energy. *Ear and Hearing* 4(4):197-201.
Mantysälo, S., and Vuori, J. 1984. Effects of impulse noise and continuous steady state noise on hearing. *British Journal of Industrial Medicine* 41:122-132.
Melnick, W. 1982. Damage risk criteria. In *Forensic Audiology*, eds. M.B. Kramer and J.M. Armbruster. Baltimore: University Park Press.
Melnick, W. 1985. Industrial hearing conservation. In *Handbook of Clinical Audiology*, 3d ed., ed. J. Katz. Baltimore: Williams and Wilkins.
Mills, J.H., Osguthorpe, J.D., Burdick, C.K., Patterson, J.H., and Mozo, B. 1983. Temporary threshold shifts produced by exposure to low frequency noises. *Journal of the Acoustical Society of America* 73(3):918-923.
Mitchell, S.A. 1984. Noise pollution in the neonatal intensive care nursery. *Seminars in Hearing* 5(1):18-24.

Moody, D.B., and Stebbins, W.C. 1986. Effects of noise exposure on absolute threshold: Experimental sensorineural changes. *Seminars in Hearing* 7(1):39-47.

National Institutes of Occupational Safety and Health (NIOSH) 1972. Criteria for a recommended standard . . . occupational exposure to noise. Report HSM 7311001. Department of Health, Education and Welfare, Washington, D.C.

Plakke, B.L. 1983. Noise levels of electronic arcade games: A potential hearing hazard to children. *Ear and Hearing* 4(4):202-203.

Sataloff, J., Sataloff, R.T., Menduke, H., Yerg, R., and Gore, P.P. 1984. Hearing loss and intermittent noise exposure. *Journal of Occupational Medicine* 26:649-656.

Saunders, J.C., Dear, S.P., and Schneider, M.E. 1985. The anatomical consequences of acoustic injury: A review and tutorial. *Journal of the Acoustical Society of America* 78(3):833-860.

Spoendlin, H. Innervation densities of the cochlea. *Acta Otolaryngologica* (Stockholm) 73:235-248.

Stephenson, M.R., Nixon, C.W., and Johnson, D.L. 1980. Identification of the minimum noise levels capable of producing an asymptotic temporary threshold shift. *Aviation and Space Environmental Medicine* 51:391-396.

Stevin, G.O. 1982. Spectral analysis of impulse noise on hearing conservation purposes. *Journal of the Acoustical Society of America* 72(6):1845-1854.

Swansow, S.J., Dengerink, H.A., Kondrick, P., and Miller, C.L. 1987. The influence of subjective factors on temporary threshold shifts after exposure to music and noise of equal energy. *Ear and Hearing* 8(5):288-291.

Ward, W.D., Santi, P.A., and Duvall, A.J. 1981. Total energy and critical intensity concepts in noise damage. *Annals of Otology, Rhinology and Laryngology* 90:584-590.

The Effects of Noise on Learning, Cognitive Development and Social Behavior

ARLINE L. BRONZAFT

Noise and Learning

Laboratory Studies. Early laboratory work (Kryter 1950, 1970; Broadbent 1957) did not find compelling support for any detrimental impact of noise on mental and psychomotor performance. Kryter (1970) had postulated that individuals adapt to noise. In his latest comprehensive review of the effects of noise on mental and psychological task performance, Kryter (1985) concludes that while noise can be detrimental to performance that involves audition, at other times it may mask competing sounds and improve performance; and that it is the attitudes of learners toward noise that may bring about a decline in performance. However, Glass and Singer (1972), after surveying existing literature on noise and conducting their own studies, reported that exposure to unpredictable and uncontrollable high-intensity noise often leads to a degradation in the quality of task performance, impaired ability to resolve cognitive conflict, and lowered tolerance for frustration.

Since many of the laboratory studies were conducted on adult subjects, it could be argued that adult and child developmental differences in the areas of perception, language, cognition, and memory (Piaget and Inhelder 1969; Gibson 1969) call for separate studies on the impact of noise on children's performance. Results of short-term studies of imposed noise on school-age children have run the gamut from no effects on written performance (Slater 1968) to adverse effects on recognition memory tasks (Wyon 1968).

Natural Settings. The 1970s saw the questioning of the applicability of findings from laboratory studies, in which individuals are typically exposed to noise for short periods of time, to situations where people are exposed for longer periods of time. However, naturalistic environmental studies are also suspect because they often lack the experimental control demanded. It was the hope of environmental psychologists, who came to the forefront in the 1970s, that the type of correlational studies they would conduct in the natural setting would provide good data from which to draw conclusions.

Cohen and his colleagues (1973), in one of the early studies on the effect of noise in a natural environment, found that elementary-school children who were living on the lower floors of buildings directly exposed to high-intensity expressway noise showed greater impairment in reading ability than did children living on higher-floor apartments. Cohen et al. (1980) followed this study with one looking at the effects of noise on children who attended elementary schools in the air corridor of the Los Angeles International Airport. Control subjects attended schools in quiet communities. Children attending the noisy schools were found more likely to fail to perform cognitive tasks and to give up before the time allotted for completion of these tasks. However, these researchers did not find that noise interfered with reading ability, and they attributed this to the fact that students in this study attended different schools (Cohen et al. 1980).

At the time Cohen was conducting his research on highway and airport noise, Bronzaft and McCarthy (1975) examined the impact of the noise of elevated trains on the reading ability of children attending a school adjacent to such train tracks in New York City. The school that served as the field for this research was set up so that half the classes faced the tracks and the other half were located on the quiet side of the building. This arrangement allowed for greater control of variables in that children on both sides of the building came from similar socioeconomic backgrounds and were being exposed to similar teaching methods. Eighty trains passed along the tracks each day between the hours of 9:00 A.M. and 3:00 P.M. when classes were in session. The prevailing noise level of a class in session was about 59 dBA, but the level rose to 89 dBA when a train passed by. Students on the noisy side did more poorly on reading achievement tests than did those on the quiet side of the building. The children in the sixth grade who were exposed to the noise were found to be a year behind their counterparts on the quiet side.

Further investigation of the effects of noise on learning was reported on by Green et al. (1982) in a study of the reading ability of children who attended schools near the noise-impact contours of the John F. Kennedy and La Guardia airports in New York City. Their findings revealed that the percentage of students reading below grade level increased as noise level increased. Lukas et al. (1981) reported that higher classroom noise levels correlated with lower reading achievement when schools nearer to and farther away from Los Angeles freeways were compared. Hambrick-Dixon (1986) tested the effects of noise on psychomotor performance in preschoolers attending day-care centers near New York City's elevated trains, and reported that such performance was indeed adversely affected.

Effects of Quieting. When the Transit Authority of New York City installed rubber vibration isolator pads on the tracks adjacent to the school cited above and the Board of Education of New York

City installed acoustic ceilings in three of its noisiest rooms facing the tracks, a study was conducted to examine the impact of these noise abatement measures (Bronzaft 1981). Reading scores for children on both noisy and quiet sides were obtained for the year prior to the noise abatement treatments, and, as expected, children on the noisy side did significantly poorer. After the noise levels on the noisier side were reduced by six to eight decibels, children on both sides of the building were found to be reading at comparable levels. While this study demonstrated the effectiveness of noise abatement, it also provided additional data to support the proposition that noise has an adverse effect on reading level.

Lehmann and Alphandery (1983) carried out another study involving acoustic modifications in a school building; in this study, children's classroom behavior was observed before and after sound insulation materials were installed. Following the noise reduction, the children's attention and active participation in classroom activities were found to be improved.

Noise and Cognitive Development

Noise has been found to take its toll on children even at a very early age. The prevalence of noise unrelated to an infant's actions and goals can result in reduced inclination to orient and attend to adult speech (Hunt 1979). Wachs (1982) noted that young children acquire language skills more slowly in noisy homes and were less likely to explore their surroundings. Children living in crowded conditions with loud voices and blaring radio or television speakers have been found to discriminate vocal sound patterns less well than do children from quieter middle-class homes (Clark and Richards 1966). Heft (1979) has reported that noisy homes adversely affect incidental memory performance in young children. In such noisy environments, children are less likely to have the interaction with adults that foster these skills. According to Wachs, Uzgiris, and Hunt (1971), the hampering effects on psychological development of slum-rearing under conditions of high noise prevalence may lie in stimulus bombardment of the child rather than in stimulus deprivation.

Evidence relating to the impact of noise on children's early language development, on psychomotor skills, and subsequently on reading skills, suggests that the above studies, although not conclusive, indicate real losses in cognitive development and learning. These studies, while correlational in nature, have been carefully conducted and should carry some weight. They have implications for the training of parents and caregivers, as well as for compensatory education (Hunt 1971). Their findings suggest very strongly that additional investigations are needed.

Noise and Social Behavior

We are all familiar with anecdotal reports of violence and aggression breaking out because a radio is being played too loudly, a party goes on past midnight, or a motorcycle race is taking place in a residential neighborhood. Laboratory studies on the relationship between aggression and noise have shown that when subjects were exposed to noise, they were more likely to administer electric shocks (no shock was actually administered, but the subject did not know this until the end of the experiment) to an ostensible victim (Geen and O'Neal 1969; Donnerstein and Wilson 1976).

Milgram (1970), in discussing the deterioration of social graces in large urban centers, pointed to the overload of stimuli as being partially responsible for such urban ills as the tendency of passersby to ignore individuals who are in need of help. While it is not specifically identified as the sole culprit, noise is seen as a major contributor to such sensory overload. In field studies, subjects have appeared less helpful, if not hostile, when exposed to the noise of a lawnmower at the time a passerby needed help with dropped books (Mathews and Cannon 1975). The noise of jackhammers also made it less likely that someone in need would receive help (Page 1977).

In 1986, the New York City Police Department's patrol cars responded to approximately 10,000 noise complaints. In that same year, the New York City Department of Environmental Protection said it received about 3,000 noise complaints; three years later, the number of complaints rose to 6,331. However, only about 20% of that department's complaints resulted in the issuance of summonses. This means that many people who complained about noise experienced little relief because of limited enforcement of the City's noise code, and one many hypothesize that there were many very angry New Yorkers as a result.

Bronzaft and Santa Maria (1989) reported on a confrontation between New York City police and rioters, with noise identified as one of the factors underlying the incident. They also describe other situations where noise may lead to increased urban violence. Apparently, New Yorkers are not the only people enraged by noise. A man in Brisbane, Australia faced a prison term for beating two dogs with a mallet when their owner refused to respond to his repeated requests to quiet them (*Courier Mail* 1987).

Conclusion

The few studies cited above, plus the many reports in the news media of fights and confrontations erupting over noise intrusions, suggest that further research in the relationship between noise and social behavior would be fruitful.

REFERENCES

BROADBENT, D.E. 1957. Effects of noise on behavior. In *Handbook of Noise Control,* ed. C. M. Harris. New York: McGraw-Hill.

BRONZAFT, A.L. 1981. The effect of a noise abatement program on reading ability. *Journal of Environmental Psychology* 1: 215-222.

BRONZAFT, A.L. and MCCARTHY, D.P. 1975. The effect of elevated train noise on reading ability. *Environment and Behavior* 7: 517-528.

BRONZAFT, A.L., and SANTA MARIA, C. 1989. Noise annoys, but it also masks crime and incites violence. *Law Enforcement News* 15: 8.

CLARK, A., and RICHARDS, C. 1966. Auditory discrimination among economically disadvantaged and non-disadvantaged preschool children. *Exceptional Children* 33:259-262.

COHEN, S., GLASS, D., and SINGER, J. 1973. Apartment noise, auditory discrimination and reading ability in children. *Journal of Experimental Social Psychology* 9: 422-437.

COHEN, S., EVANS, G.W., KRANTZ, D.S., and STOKOLS, D. 1980. Physiological, motivational and cognitive effects of aircraft noise on children. *American Psychologist* 35: 231-243.

Courier Mail, April 2, 1987. Committed for trial. p. 2.

DONNERSTEIN, E., and WILSON, D.W. 1976. Effects of noise and perceived control on ongoing and subsequent aggressive behavior. *Journal of Personality and Social Psychology* 34: 774-781.

GEEN, R.G., and O'NEAL, E.C. 1969. Activation of cue-elicited aggression on general arousal. *Journal of Personality and Social Psychology* 11: 289-292.

GIBSON, E.J. 1969. *Principles of Perceptual Learning and Development.* New York: Appleton-Century-Crofts.

GLASS, D.C., and SINGER, J.E. 1972. *Urban Stress: Experiments on Noise and Social Stressors.* New York: Academic Press.

GREEN, K.B., PASTERNAK, B.S., and SHORE, R.E. 1982. Effects of aircraft noise on reading ability of school-age children. *Archives of Environmental Health* 37: 24-31.

HAMBRICK-DIXON, P.J. 1986. Effects of experimentally imposed noise on task performance of black children attending day care centers near elevated subway trains. *Developmental Psychology* 22: 259-264.

HEFT, J. 1979. Background and local environmental conditions of the home and attention in your children. *Journal of Applied Social Psychology* 9: 47-69.

HUNT, J. McV. 1979. Developmental psychology: Early experience. *Annual Review of Psychology* 30:103-143.

KRYTER, K.D. 1950. Noise and behavior. *Journal of Speech and Hearing Disorders* 15: Monograph Supplement 1.

KRYTER, K.D. 1970. *The Effects of Noise on Man.* New York: Academic Press.

KRYTER, K.D. 1985. *The Effects of Noise on Man.* 2d ed. Orlando: Academic Press.

LEHMANN, A., and ALPHANDERY, H.G. 1983. Effect of noise on children at school. In *Proceedings of the Fourth International Congress on Noise as a Public Health*

Problem, Vol. II, ed. G. Rossi. Milano: Edizioni Tecniche a curo del Centro Ricerche e Studi Amplifon.

LUKAS, J.S., DUPREE, R.B., and SWING, J.W. 1981. *Effects of Noise on Academic Achievement and Classroom Behavior*. Berkeley, California: Office of Noise Control, Department of Health Services.

MATHEWS, K.E., and CANON, L.K. 1975. Environmental noise level as a determinant of helping behavior. *Journal of Personality and Social Psychology* 32: 571-577.

MILGRAM, S. 1970. The experience of living in cities. *Science* 167: 1461-1468.

PAGE, R.A. 1977. Noise and helping behavior. *Environment and Behavior* 9: 559-572.

PIAGET, J., and INHELDER, B. 1969. *The Psychology of the Child*. New York: Basic Books.

SLATER, B. 1968. Effects of noise on pupil performance. *Journal of Educational Psychology* 59: 239-243.

WACHS, T.D. 1982. Relation of home noise-confusion to infant cognitive development. Paper presented at the Annual Meeting of the American Psychological Association, Washington, D. C.

WACHS, T.D., VZGIRIS, I.C., and HUNT, J. McV. 1971. Cognitive development in infants of different age levels and from different environmental backgrounds: an explanatory investigation. *Merrill-Palmer Quarterly of Behavior and Development* 17(4):283-317.

WYON, D. 1968. Studies of children under imposed noise and heat stress. *Ergonomics* 13: 598-612.

Community Response and Attitudes Toward Noise

ARLINE L. BRONZAFT AND JANE R. MADELL

Noise and Annoyance

The human response to unwanted sound involves far more than just the simple assessment of its physical intensity, the perception of which is called its loudness. Other factors can affect the undesirability of a sound, such as its unique combinations of frequencies and temporal patterns as they interact with its intensity. Furthermore, the attitude of a given listener toward a particular sound, its source, or its associated meaning can strongly affect that listener's reaction to it. For instance, if an individual does not enjoy listening to a particular style of music, hearing it at even a relatively reduced level can cause a negative reaction. And if the listener is unable to control the source, or at least escape the sound, simple annoyance can escalate into anger, rage, or even violence.

Annoyance is a common judgmental response to noise regardless of its level. According to Miller (1974), it "has its base in the unpleasant nature of some sounds, in the activities that are disturbed or disrupted by noise, in the physiological reactions to noise, and in the responses to the meaning or messages carried by the noise."

Identifying the Factors of Annoyance

Borsky (1969) identified the following factors as affecting the acceptability of certain sounds:

1. Feelings about the necessity or preventability of sound can determine its acceptability. When listeners feel that the propagators of an intruding sound are callous and indifferent to their needs, the sound is more likely to be annoying, even at a fairly low level.
2. Feelings about the value of a sound source's primary function have a significant effect. Sound caused by building a home is more likely to be acceptable than sound caused by screeching tires.
3. The type of living activities affected determines how annoying a sound will be. It is more difficult to accommodate to sound that interferes with sleep or relaxation than to sound that may be present during ordinary waking activities.

4. People's satisfaction with their living environment will affect the acceptability of certain sound sources. If other things in their environment are distressing, increased levels of noise will be more difficult to tolerate.

5. People who believe that noise has a negative effect on health will find noise more annoying than those who do not.

6. The liability to feel annoyance with noise exhibits individual differences.

7. The relationship between fear and sound is a significant factor. Response to aircraft noise, for one example, may be significantly affected by fear of crashes. Fear of the sound of screeching tires, for another, may be worse to parents when their teenage child is out driving than at other times. When fear is not a factor, an individual can learn to put a sound in the background.

8. Several other factors affect the acceptability of sound. These include the extent of positive feelings about the area in which a person lives, whether a residence is considered permanent or temporary, and past experiences with sound.

Borsky did not find age, sex, social class, family composition, or length of residence to correlate significantly with annoyance (1969). He had earlier reported that, despite having adjusted to the fact that noise is to be present, people's feeling of annoyance over noise interference does not show a habituation effect, but remains the same, or even grows over the years as they continue to live in a noisy area (Borsky 1961).

Cohen (1969) reported that sounds of higher intensity or higher frequency, and sounds that are variable in nature rather than steady, are more often rated as annoying. Annoyance does not begin at some absolute exposure level, but rather at a relative exposure level. However, once that level is reached, annoyance may grow quite rapidly (Fidell et al. 1983). And, sensitivity to noise appears to be higher at those times when it interferes with sleep (Gyr and Grandjean 1984).

The Relation of Noise Annoyance to Feelings and Attitudes

Sounds can influence our attitudes because of the information they convey. When unpleasant information is being transmitted, these sounds may be perceived as annoying. Thus, certain sounds may be judged annoying because they convey alarm or distress rather than because of their acoustic properties (Cohen 1969). Fire engine sirens can elicit annoyance reactions because of the fear associated with fires. Similarly, the sound of approaching aircraft can be stressful because it elicits fear of a plane crash. This is particularly significant in communities near airports. It has also been reported (Broadbent 1980) that complaints about aircraft noise come disproportionately from people who are afraid to fly. Hall (1984) has reported that aircraft noise was found to be more annoying than road noise, which, in turn, was found to be significantly more annoying than rail noise. Furthermore, when people think that it is easy to reduce traffic noise levels or that the vehicles producing it are not particularly necessary, they are more annoyed by the traffic noise (Bradley 1980).

It might be expected that people exhibit less annoyance over time, but Weinstein (1982), examining the effects of adaptation to traffic noise after the opening of a highway, found that there was no adaptation in his subjects in self-reported annoyances. Over time, Weinstein's subjects became more pessimistic about their ability to adapt to the noise.

Kryter (1980) contends that it is our negative feelings about the interference of noise with our normal everyday life, rather than the noise itself, that may bring about ill health effects.

Annoyance Assessment and Community Response

One way to study the relationship between annoyance and community reactions is through social surveys. Miller (1974) points out, however, that there are difficulties with these social surveys in that the mere presence of observers in the community, the method of the interview, and the construction of the questionnaire may influence the data collected. Annoyance has also been found to be related to certain personality factors, such as depression and hysteria as measured by the MMPI (Francois 1980), a fact necessitating collection of data from large samples. Kryter (1985) notes that attitude ratings are also influenced by the activities that individuals are engaged in at the time of interviews; thus, several assessments are required to help ensure reliability. Still, Kryter concludes that "the reliability of attitude survey data appears to be greater than originally thought." Most of this research has focused on the impact of noise from transportation vehicles, with aircraft noise being the most disturbing, and street-traffic noise being the second most upsetting.

There have been too few studies of noise annoyances within other settings such as the workplace, the school, the hospital, and the home. Recognizing that much of the data on community response to noise exposure concerned aircraft noise, the U. S. Environmental Protection Agency (1977) carried out a study published as *The Urban Noise Survey.* This national survey looked at community reaction over a broad range of noise conditions. The survey found that individuals living in urban communities not exposed to aircraft or highway noise were still annoyed by transportation noise, and were also disturbed by the noise of neighbors, pets, construction, power garden equipment, radios, and televisions. Widespread annoyance was expressed by the individuals surveyed, who complained that noise degraded the quality of life in their urban environments.

Aitkin (1982) had found through quantitative noise analyses that modern hospitals were noisier than they should be, confirming anecdotal comments from patients that they could not get much rest during their hospital stays. Schoolteachers and administrators (Bronzaft 1988) have long complained about noise in their classrooms and elsewhere in their school buildings. Langdon, Buller, and Scholes (1983) have shown that noise caused by neighbors is particularly disturbing. While it is true that individuals expect some noise to emanate from nearby neighbors, especially in large apartment houses, they also feel that there are too many neighbors who take advantage of this expectation. Workers in an occupational setting such as a wheel and axle shop might reasonably expect their working environment to be excessively noisy, but many workers in business offices with large spaces divided by inadequate partitions have also come to view their surroundings as excessively noisy as well.

Annoyance and Community Complaints

Borsky (1980) reported that, despite the fact that very large numbers of community residents are disturbed by environmental noise, little has been done to abate the nuisance. One reason is that too few people actually complain to the appropriate officials about these annoyances. Borsky (1969), in an earlier paper, reported that research in both the United States and England had indicated that only about 20 to 23 percent of the exposed population complained about intruding noises. Factors influencing the likelihood of complaints included degree of annoyance, the belief that complaining offered the possibility of making a change, the seriousness of the noise problem in comparison to other problems in the area, and the availability of organizations in the area that dealt with noise complaints. In his later review (1980), Borsky still found relatively little complaint behavior. Even those people who were especially annoyed and had organized themselves believed their chances of successfully abating community noise were low.

Kryter (1985) also finds that complaints and legal or political actions are not good measures of community annoyance because there are too few of them. However, in the time since Kryter's publication, there have been growing numbers of complaints from community groups, some of which are determined to achieve success in lowering the decibel level in their communities. Ruben (1991) has reported that in New Jersey alone there are forty-one groups working to curtail airport noise and that there are twenty-nine such groups in California. These groups are working together with legislators to have aircraft noise officially recognized as a public problem, and they have pledged to persist in their efforts. The Port Authority of New York and New Jersey issued a press release (1991) describing their school soundproofing program for the year. Thirty-four schools are included in their soundproofing efforts, which have reduced perceived noise levels in the classrooms by at least half. Credit for prevailing upon the Port Authority to authorize this noise abatement project must be given to the communities that had long lobbied for this program.

Robinson (1991) reports that local communities are enacting broader ordinances "aimed at a wide range of noise pollution." The residents of Burlington, Vermont, will receive fines if they disturb the community peace with their cars or stereos.

Citizen groups with strong leadership can bring about significant change when they bring their collective complaints to bear on specific noise sources. One major success story is that of The Big Screechers in New York City, a large constituency made up of several thousand citizens annoyed by public transit noise generated by subway and elevated trains. Founded and held together essentially by the persistence of one strong citizen leader, Carmine Santa Maria, with the assistance and support of a small group of scientists and environmentalists at its core, the group has successfully forced the enactment of a New York State transit noise code that specifies noise abatement goals, maximum permissible noise emission levels, and compliance dates, and that calls for annual published progress reports. Another significant accomplishment of The Big Screechers was to bring about the fitting of new and existing Metropolitan Transportation Authority subway cars with ring-damped wheels that reduce their screeching as they round curves or brake. Ten years of collective citizen effort has resulted in the substantial mitigation of a perennial source of annoyance for hundreds of thousands of daily riders. Present and future generations of New Yorkers will owe this energetic group of citizens a long-standing debt of appreciation. Other groups can and will follow suit, but they must be prepared to work hard and long to achieve similar goals. Nevertheless, the reward of widespread reduction in public annoyance and hazards to hearing health is well worth the effort.

Suggestions for Lessening Noise Annoyance

Annoyance brought about by noise may lead to psychological stress if the noise either intrudes upon ongoing activities or carries disturbing messages. However, if noises are accompanied by tangible benefits and disturb residents for only limited amounts of time, as in the case of some necessary construction projects, community residents, when informed of the goals of the projects and the length of time they will have to endure the noises, may experience them as more readily tolerable. Thus they may feel that they have some control over their lives. It is when people lack such control that they feel the sort of learned helplessness that can produce emotional disturbances (Seligman 1975).

When individuals believe that the offending transportation agencies are moving ahead speedily in lowering the noise levels of overhead aircraft and nearby elevated trains, they may be better able to cope with the noises imposed upon them by these transit vehicles. Transit agencies that understand community problems, respect residents' feelings, and are genuinely committed to noise abatement may be viewed more positively by citizens, who may consequently complain less about the intrusive noises.

One of the frequent complaints of urban dwellers concerns noises originating in their neighbors' apartments. Apartment houses could pass house rules that stipulate conditions under which certain noises will be permitted: for example, noisy construction and repair might be allowed on weekdays from 9 A.M. to 4 P.M. The quality of life for urban dwellers would undoubtedly be enhanced by compliance with such rules. Furthermore, if residents occasionally overstep the house rules, for instance by having a noisy party, complaints may be prevented if they inform their neighbors ahead of time, let their neighbors know they are, say, celebrating something special, and will end the party at a reasonable hour. Where there is neighborly respect and consideration, there should be less annoyance.

These suggestions for lessening annoyance caused by various sources and types of noise are extrapolated from the work and observations of Seligman (1975). There is, however, little or no documented research reporting the success or failure of efforts to lessen noise-induced annoyance, and this is an area that calls for substantial study.

REFERENCES

AITKEN, R.J. 1982. Quantitative noise analysis in a modern hospital. *Archives of Environmental Studies* 37(6):361-364.

BORSKY, P.N. 1961. Community Reactions to Air Force Noise. Parts I and II. WADD Technical Report 60-689 (I) and (II), U.S. Air Force.

BORSKY, P.N. 1969. Effects of noise on community behavior. In *Proceedings of the Conference on Noise as a Public Health Hazard* (Washington), *ASHA Reports 4*, eds. W.D. Ward and J.E. Fricke. Washington, D.C.: American Speech and Hearing Association.

BORSKY, P.N. 1980. Review of community response to noise. In *Proceedings of the Third International Congress on Noise as a Public Health Hazard* (Freiburg), *ASHA Reports 10,* eds. J. Tobias, G. Jansen, and W.D. Ward. Rockville, Maryland: American Speech-Language-Hearing Association.

BRADLEY, J.S. 1980. Field study of adverse effects of traffic noise. In *Proceedings of the Third International Congress on Noise as a Public Health Hazard* (Freiburg), *ASHA Reports 10,* eds. J. Tobias, G. Jansen, and W.D. Ward. Rockville, Maryland: American Speech-Language-Hearing Association.

BROADBENT, D.E. 1980. Noise in relation to annoyance, performance, and mental health. *Journal of the Acoustical Society of America* 68(1):15-17.

BRONZAFT, A.L. 1988. The hazards of noise. In *Teacher Matters,* Vol. 4 (2). New York: N.Y.C. Teacher Centers Consortium.

COHEN, A. 1969. Effects of noise on psychological state. In *Proceedings of the Conference on Noise as a Public Health Hazard* (Washington), *ASHA Reports 4,* eds. W.D. Ward and J.E. Fricke. Washington, D.C.: American Speech and Hearing Association.

FIDELL, F.S., HORONJEFF, R., SCHULTZ, T., and TEFFETELLER, S. 1983. Community response to blasting. *Journal of the Acoustical Society of America* 74(3):888-893.

FRANCOIS, J. 1980. Aircraft noise, annoyance, and personal characteristics. In *Proceedings of the Third International Congress on Noise as a Public Health Hazard* (Freiburg), *ASHA Reports 10,* eds. J. Tobias, G. Jansen, and W.D. Ward. Rockville, Maryland: American Speech-Language-Hearing Association.

GYR, S., and GRANDJEAN, E. 1984. Industrial noise in residential areas: Effects on residents. *International Archives of Occupational and Environmental Health* 53(3):219-231.

HALL, F.L. 1984. Community response to noise: Is all noise the same? *Journal of the Acoustical Society of America* 76(4):1161-68.

KRYTER, K.D. 1980. Physiological acoustics and health. *Journal of the Acoustical Society of America* 68(1):10-17.

KRYTER, K.D. 1985. *The Effects of Noise on Man.* 2d ed. Orlando: Academic Press.

LANGDON, F.J., BULLER, I.B., and SCHOLES, W.E. 1983. Noise from neighbors and the sound insulation of party floors and walls in flats. *Journal of Sound and Vibration* 88:243-270.

MILLER, J.D. 1974. Effects of noise on people. *Journal of the Acoustical Society of America* 56(3): 729-764.

Port Authority of New York and New Jersey. 1991. Press release, May 16.

ROBINSON, M. 1991. What? *The Richmond News Leader* (People Section), May 3. Richmond, Virginia.

RUBEN, B. 1991. On deaf ears. *Environmental Action* March/April:16-19.

SELIGMAN, M.E.P. 1975. *Helplessness: On Depression, Development, and Death.* San Francisco: W.H. Freeman.

U.S. Environmental Protection Agency, Office of Noise Abatement and Control. 1977. *The Urban Noise Survey.* Washington, D.C.: Contract No. 68-01-4184.

WEINSTEIN, N.D. 1982. Community noise problems: Evidence against adaptation. *Journal of Environmental Psychology* 2:99-108.

Noise Abatement

SAMUEL STEMPLER

If efforts to abate the crescendo of noise now assaulting our communities are to be successful, our legislators must be convinced that cost-effective strategies are available and will be accepted by their constituents.

STRATEGIES

Noise abatement requires active participation and involvement on two levels. First, where individuals have direct personal control of the source of noise or of the manner of its reception, they can take actions of their own to mitigate the harmful and annoying effects of noise. And second, for those sources over which little or no personal control can be exerted, governmental agencies should be pressed to enforce and strengthen noise control legislation.

Personal Involvement. On the personal level there is much that can be done:

1. Insulate the home.
2. Demand less noisy equipment from manufacturers.
3. Purchase only low-noise products.
4. Get involved. Make noise about noise to abate noise. Seek out and join forces with others who are concerned about noise. Groups of constituents can bring about change more effectively than individuals. Legislators serve best those constituents who make their concerns known.

Government Action. Regulatory efforts to control noise are mounted at all three levels of government: Federal, state, and local. Unfortunately, on the Federal level, authority and responsibility to control noise are scattered among several agencies.

The Department of Transportation (DOT) establishes criteria and sets standards to control aircraft noise through the Federal Aviation Administration (FAA), and highway noise through the Federal Highway Administration (FHA). The Occupational Safety and Health Administration (OSHA), part of the Department of Labor, sets and enforces noise standards in the workplace. The Department of Housing and Urban Development (HUD) has developed noise criteria for housing. Many other Federal agencies also have a role in noise control and abatement. Unfortunately, these important concerns are not the prime agenda and responsibility of any of these agencies. On the contrary, noise control is often perceived as a hindrance. For example; during the closing days of the 101st Congress, air-traffic legislation was passed that severely limited noise control actions that can be taken by local agencies.

The Environmental Protection Agency (EPA) has prime responsibility for most other Federal efforts to control noise. Yet, the EPA's Office of Noise Abatement and Control (ONAC) was phased out in 1981 without resources having been provided to enforce the rules and regulations that ONAC promulgated. While the EPA took the position that noise was primarily a local problem, they left their regulations in force, without any enforcement staff, to preempt local jurisdiction. One result of this lack of support is evidenced by the fact that, of the approximately 1,100 local and state noise control programs in place in 1982, fewer than 20 still existed as of 1990 (ASHA 1990). Yet local agencies are still able to do much to control noise, even though they lack jurisdiction over such major noise sources as aircraft and railroads because of inter-state commerce considerations.

The New York City Noise Control Code (1972) illustrates many of the strategies that are appropriate at the municipal level. Examples of some of these include the following:

1. The code prohibits some activities without regard to noise level. For example, the use of a public address system to attract attention to a business is altogether prohibited.
2. The code recognizes that certain activities are necessary, but it limits the noise level that is allowed. For example, music of any kind originating from a commercial establishment may not create noise levels in a dwelling unit greater than 45 dBA or 45 dB in any one-third octave band below 500 Hz.
3. Curfews limit construction activities to weekdays between the hours of 7 A.M. and 6 P.M., except in cases of bona fide emergencies, for which special permits are required.
4. Ambient noise standards for various land uses—residential, commercial, or manufacturing—provide guidance for developers and builders.

Economic Incentives

Rules and regulations are more effective when accompanied by an economic incentive, as witness the reduced beverage container litter and reduced burden on landfills that resulted after the imposition of a deposit refundable upon return of the container. In the case of noise, concert halls in Boise, Idaho and in London require payment of additional fees by performers who produce sound levels in excess of 85 dBA. Fees increase as sound levels or their durations increase beyond the specified limits.

Air Traffic. Aircraft pay landing fees to use airports. Several countries (e.g., France, Japan, the United Kingdom, and Germany) have added to their landing fee an additional charge scaled to the average noise generated by the aircraft (Alexandre and Barde 1987a). These fees serve several purposes. They provide encouragement to airlines to retrofit existing aircraft or to replace them with quieter versions, and to manufacturers to market quieter aircraft. They also provide funds to purchase or sound-treat homes and other noise-impacted facilities. Alexandre and Barde (1987a) have reported that in the Netherlands during 1985, aircraft landing noise charges were:

Aircraft	Dollars
Boeing 707	192
Boeing 727	85
DC10	56
Airbus	15
Boeing 747	15

The concept of noise-adjusted landing fees has been considered by the Federal Aviation Administration (FAA 1989) with the recommendation that "the dollar value of the noise landing fee should be set at a level that would not prohibit the continued use of noisier aircraft, but would be high enough to encourage their accelerated replacement."

According to studies by the FAA (1989), older jet aircraft, such as the Boeing 727-200, create a negative noise impact on areas surrounding an airport that are twenty times larger than those areas impacted by newer aircraft, such as the Boeing 757-200. Therefore, it is desirable to encourage early retirement, or at least retrofitting, of the older, noisier aircraft.

Vehicular Traffic. In view of the variety of ways that people operate different types of vehicles under various traffic conditions, it is probably not feasible to impose similar noise-related fees on their operators as an incentive to purchase quieter vehicles. However, such a fee, based upon relative noise level, could be imposed on all new motor vehicles sold in the country. Imposition of an average fee of two dollars would raise more than $25,000,000 a year to finance research on the health and economic costs of vehicular noise and the development of technology to reduce it.

Effect of Noise on Residential Property Values

Several studies have shown that it is the population at large, and not those who generate the noise, that pays the monetary penalty due to excessive product noise.

Alexandre and Barde (1987a) believe that cost-benefit analysis of noise reduction is hard to carry out because of the difficulty of quantifying its benefits, for example, reductions in such effects as sleep disturbance, annoyance, hearing loss, and speech interference. One possible method that they recommend to quantify the cost of excessive noise is to determine the extent to which noise causes measurable depreciation in a commodity. They suggest two approaches: one is to determine the cost of protection against noise, and the other is to determine the impact of noise upon real-estate values.

Air Traffic. In a study conducted by Starkie and Johnson (1975), it was determined, on the basis of empirical results, that people living in England near Heathrow Airport would be willing, even at a cost of 5 percent of income, to insulate their homes in order to provide a 14 dBA reduction in noise. However, this noise control strategy does not compensate for the loss of enjoyment of a balcony, garden or other outdoor amenity.

Depreciation of real-estate values because of aircraft noise was found to be in the range of 0.5 to 1.0 percent per dB of noise increase (Alexandre and Barde 1987a). On the basis of survey results, Nelson (1980) observed that a value of 0.5 percent per dB of noise may be an appropriate depreciation index for increased noise.

Newman and Beattie (1981), in their survey for the FAA, drew similar conclusions. They show that a 1 dB increase in cumulative day-night level (L_{dn})[a] noise exposure due to airport operations results in a 0.5 to 2.0 percent decrease in real-estate values, with a mean decrease of 0.7 percent.

Vehicular Traffic. Nelson (1982) studied the effects of highway noise on several indices of property value depreciation drawn from many sources, and reported that the majority of the indices reflected depreciation in the range of 0.4 to 0.5 percent. Many variables, some difficult to quantify, preclude using these indices as absolute indicators. However, as Alexandre and Barde (1987a) point out, a depreciation index of 0.5 percent per dB of noise increase is a useful guide for action when the noise is above a certain threshold, typically an equivalent level (L_{eq}).[a]

It should be noted that those communities that depend upon real-estate taxes to provide services must raise tax rates to compensate for the negative effect of revenue lost because of noise-related depreciation. While it is not clear just what this negative effect actu-

[a] L_{eq} is an equivalent steady-state sound level that contains the same acoustic energy as a time-varying sound level measured over a given period. L_{dn} is a subset of the L_{eq} measurement to which 10 dB has been added to levels measured between 10 P.M. and the following 7 A.M. in order to reflect more realistically the annoyance resulting from noise experienced during these normal hours of sleep.

ally amounts to, or how it may be validly calculated, the U.S. Department of Transportation (DOT) has addressed this country-wide problem. A study of noise pollution effects on residential property values was carried out and summarized in a DOT report by Nelson (1975). The essential results are that property values are sensitive to the difference between the A-weighted L_{10} and L_{90}[b] sound pressure levels and that the L_{10}-L_{90} difference has been shown to correlate well with annoyance (Schultz 1972).

The DOT model indicates that damages due to aircraft noise amount to $130 per average residential property per unit increase in Noise Exposure Forecast (NEF),[c] or approximately 0.5 percent per residential property per NEF. Traffic noise pollution results in damages of about $60 per dBA increase in high noise levels. This can be interpreted as having a negative effect on property values when the L_{10} levels are greater than 5 dBA above the L_{90} levels.

In 1975, these data were used to extrapolate the results of the above DOT survey in order to estimate the potential effect of noise on property values in New York City (Low et al. 1975). Using L_{10} = 65 dBA and L_{90} = 50 dBA for traffic noise, and considering aircraft noise impact upon residential property lying within the 30 Noise Exposure Forecast (NEF) contours, Low and his colleagues calculated and summarized the degradation of residential property values due to noise pollution in New York City; the results are shown in TABLE I. They caution that the results of their analysis be regarded as tentative and subject to the limitations of the data, the model specifications, and the many arbitrary assumptions needed to compute the aggregate costs, as well as their extrapolation to New York City. The conclusions should be regarded as only reasonable estimates of the true magnitudes involved.

LEGISLATION

The major sources of noise pollution, specifically, aircraft, motor vehicles, and rail transportation, can be effectively regulated only by the Federal government. The high point of the Federal Environmental Protection Administration (EPA) involvement was capped with the publication in 1977 of their document *Toward a National Strategy For Noise Control* (1977). The proposed strategies were excellent; only EPA's will to act was deficient. TABLE II, which is taken from this document, lists those EPA regulations that were actually completed and those that

[b] L_{10} is that noise level exceeded 10 percent of the time and L_{90} is that level exceeded 90 percent of the time.

[c] Noise Exposure Forecast (NEF) is a means of rating the noise impact of aircraft operations on a community near an airport. It is usually displayed as topographic contours that encompass the impacted areas.

TABLE I. Degradation of Residential Property Values Due to Noise Pollution in New York City (1975)

Class of Property	Type of Noise	Cost Per Residential or Housing Unit	Total Cost
1. One-family homes	Vehicular	$600	$233,400,000
2. All residential homes or housing units exclusive of one-family homes	Vehicular	$480	$172,274,000
3. One-family residential properties	Jet aircraft at JFK airport	$749	$50,432,050
4. All residential properties or housing units	Jet aircraft at JFK and LaGuardia airports	$650	$476,190,000

NOTE: The total degradation in residential property valuation due to vehicular traffic noise in New York City and to jet aircraft noise at JFK and LaGuardia Airports is approximately $882 million (sum of items 1, 2 and 4 above) as a first-order approximation.

TABLE II. Status of Federal EPA New Product Regulations as of 1977

	Formal Regulatory Action Begun
Completed as of 1977	Interstate motor carriers Interstate rail carriers Portable air compressors Medium and heavy trucks
Proposal to be published in Spring 1978[a]	Motorcycles Buses Truck-mounted solid waste compactors Truck-mounted refrigeration units Wheeled and crawler loaders Wheeled and crawler dozers
	Products Being Considered for Initiation of Standard-Setting Process in Near Future
If initiated, proposal would be published by early 1979[a]	Automobiles and light trucks Tires Pavement breakers and rock drills Powered lawnmowers Chain saws Guided mass transit equipment Earthmoving equipment

[a] It usually takes approximately 12 months for the final regulation to be promulgated after the publication of the proposal in the *Federal Register.* The actual effective date for industry compliance usually falls a year or more after the promulgation of the final standard.

were to be initiated. By 1977, only four regulations had been completed. Some were to be published "by the spring of 1978," and the rest, "if initiated, would be published not later than early 1979." As of 1991, when the need for such regulations is greater than ever, no new regulations for noise-producing products exist. Only a single regulation for product-noise labeling and one for hearing protectors were added, and those were completed prior to 1982.

Inasmuch as the Federal government has effectively withdrawn from noise abatement activities, and even though local agencies possess only limited authority, several of the strategies discussed above can be implemented to address local issues, and regional coalitions should be formed to solicit support for Federal action on the others.

Legislation at the state and local level will be approved if it includes actual standards and is enforceable by reference to measurements that are not too complex to be made in the field. Above all, there must be public support for the legislation.

For an excellent and comprehensive account of noise control and the law in the United States, the United Kingdom, and the European

Economic Community, the reader is referred to *The Noise Handbook,* edited by Tempest (1985).

REFERENCES

ALEXANDRE, A., and BARDE, J.P. 1987a. Economic instruments for transport noise abatement, ch. 24. In *Transportation Noise Reference Book,* ed. P.M. Nelson. London: Butterworths.

ALEXANDRE, A., and BARDE, J.P. 1987b. The valuation of noise, ch. 23, In *Transportation Noise Reference Book,* ed. P.M. Nelson. London: Butterworths.

ASHA (American Speech-Language-Hearing Association) 1990. *Noise Regulation Report* 17(1):1.

FAA (Federal Aviation Administration) 1989. *Report To Congress. Status of U.S. Stage 2 Commercial Aircraft Fleet* Required by House Report No. 101-691, Washington, D.C.

LOW, R., ELDON, E., STEMPLER, S., WATKINS, H., SANDERS, H., and BORONOW, E. 1975. *Ambient Noise Quality Zones, Criteria, and Standards.* New York: A report to the City Council.

NELSON, J.P. 1975. *The Effect of Mobile-Source Air and Noise Pollution on Residential Property Values,* 75–76. Department of Transportation DOT-TST.

NELSON, J.P. 1980. Airports and property values: A survey of recent evidence. *Journal of Transport Economics and Policy* 14:37-52.

NELSON, J.P. 1982. Highway noise and property values: A survey of recent evidence. *Journal of Transport Economics and Policy* 16:117-138.

NEWMAN, J.S., and BEATTIE, K.R., March, 1981. *Aviation Noise Effects.* Federal Aviation Administration Report No. FAA-EE-85-2.

New York City Noise Control Code 1972. In *Administrative Code of the City of New York* Title 4, Chapter 2.

SCHULTZ, T. 1972. *Community Noise Ratings.* London: Applied Science Publishers Ltd.

STARKIE, D., and JOHNSON, D. 1975. *The Economic Value of Peace and Quiet.* Lexington: Saxon House.

TEMPEST, W. (ed.). 1985. *The Noise Handbook.* Orlando: Academic Press.

Toward A National Strategy For Noise Control 1977. Washington, D.C.: Office of Noise Abatement and Control, Environmental Protection Administration.

Public Education and Awareness of the Effects of Noise

ARLINE L. BRONZAFT

Government and Community Responsibilities

In the 1970s the Federal government published excellent educational materials to alert local authorities to the dangers of noise, assisted these local governments in establishing their own noise control programs, and urged local communities to pass noise control ordinances. Regrettably, the early 1980s saw the virtual demise of the Federal effort in noise control and the task of noise abatement was left to local governments, which have generally given it low priority. Local governments have not become involved in noise research nor have they initiated programs to educate the public on the health hazards of noise. Even in those areas where there are anti-noise ordinances, they often go unenforced. Thus, if we are to be successful in combating the harmful effects of noise, concerned citizen groups will need to assume a major role in this effort.

Citizens can lobby to reinstate Federal assistance programs for noise control and they can also pressure their own local governments to place a higher priority on noise abatement. By joining together in anti-noise organizations, citizens can become effective advocates for quieter and healthier communities and can serve as valuable allies to schoolteachers in their efforts to inform their students of the hazards of noise.

Awareness in the Schools

Elementary Schools. Efforts to raise the consciousness of the general public on a fundamental health issue such as the hazards of noise should start with the education of young children as early as kindergarten and elementary school, when they are still impressionable. In the case of exposure to noise, children of elementary-school age are probably exposed more to recreational and household noises than to other noise sources. Some, however, assist their parents in noisy work before and after school, in such tasks as farming, dairying, construction, and those involving the use of power tools. These children may be exposed to long periods of excessive machinery or engine noise. They and their families need to be made aware of the special hazards to health posed by such exposures and of means by which to protect themselves. Efforts to reach such children should be made a part of the general health education curriculum so that all children would be alerted to the hazards of noise whether encountered at work or during recreational activities. Available teaching materials are generally inadequate to the needs of teachers (Frager and Khan 1988). Many have expressed a desire for assistance, including curriculum guidance and teaching materials, as well as direct assistance from various specialists with relevant expertise, such as audiologists, speech-language pathologists, psychologists, industrial hygienists, engineers, and other professionals who have developed an interest in the problems of noise. (Several organizations listed later in this section could be helpful in supplying some of these needs.)

High Schools and Vocational Schools. Florentine (1990) has reported that the inadequacy of hearing-health education in junior and senior high schools has been documented in surveys that reveal the extent of misinformation among students about hearing damage due to noise and the prospects of recovery from it (Lass et al. 1987; Lewis 1989). Prevention of hearing damage before it occurs is the central thrust of Roeser's "Protect the ear before the 12th year" (1980); and Arthur (1988) has pointed out that, while hearing assessment programs are well-established in the schools, there are few hearing conservation programs with effective follow-up activities. These are especially needed in the vocational and technical schools where students are liable to be exposed to hazardous noise levels in workshop training sessions (Florentine 1990). Suter (1986) has described seven elements of an effective hearing conservation program: noise measurement, engineeringand administrative controls, audiometric testing, hearing protection, education, record keeping, and program evaluation.Florentine (1990) points out that Melnick (1984) has urged the use of evaluations of program effectiveness to ensure their ultimate success. Killip and associates (1987) have noted the importance of integrated school and community programs to the enrichment of the education of both students and their parents to the hazards of noise, particularly from recreational sources.

Community Support Groups. It was partly in this spirit that the Noise Committee of the Council on the Environment of New York City was formed, and it could serve as a model for the formation of similar community education support groups.

The Council on the Environment of New York City is a privately funded citizens' organization in the Office of the Mayor. The Council engages in several environmental activities, including the operation of an open-space greening program, the sponsoring of an award-winning Greenmarket program consisting of a series of farmers markets throughout the city, the running of an office-paper recycling service, and the management of an educational program that trains student organizers. The Council also has several committees, including a noise committee, that are composed of volunteer citizens, most of whom are environmental experts. Members of the noise committee are frequently consulted by all branches of the media for their professional expertise, and they serve as a major international source of public information on noise.

As part of the noise committee's efforts to educate youngsters about the hazards of noise and provide them with specific strategies to combat noise in the environment, the Council provided the New York City Board of Education with 6,000 posters highlighting the dangers of noise to physical and mental health and suggesting ways to quiet our noisy surroundings. These posters, together with pamphlets entitled "Noise is more than a pain in the ear. . .," were introduced to all elementary schools in the fall of a recent year. To assist teachers with their lectures on the dangers of noise, the Council also provided them with a sample noise lesson that could be adapted to different grade levels.

This noise lesson introduces students to the sources of noise in the urban environment, describes briefly how noise is measured, and how it affects hearing. The physiological and psychological effects of sustained noise exposure are also discussed. Students may thus gain some familiarity with the City's noise code and be encouraged to abide by its regulations. The agencies involved in combating noise pollution are identified, and ways to reduce noises in the home and the school are also to be presented.

One of New York City's school districts recently took advantage of the Council's noise program under the guidance of a professor from the City University of New York who prepared her students to give noise lectures in the schools of that district. After the lectures, which were given in the school assembly halls, the students in that school district were given the opportunity to enter a poster and essay contest highlighting the dangers of noise. The posters and essays were then judged by a committee including the director of New York City's Bureau of Noise Abatement. Certificates were issued to the student winners.

This example of how one school district enlightened its students on the hazards of noise could be repeated, adapted as needed, in school districts throughout the city and elsewhere in the nation, provided that the administrators in charge deem the effort worthwhile.

Awareness beyond the Classroom

If school children can be encouraged to share what they have learned about noise with their families and friends, this extended group could become better equipped to work toward effecting significant change in both the home and the community at large. Downs (1990) has created a fascinating program for the youth of Denver called Johnny Apple Ear that promotes hearing conservation and the use of hearing protection by youngsters. It is based on the folk character Johnny Appleseed, and has captured the imagination of many of that city's young people.

Literature. One means of effecting change has been the production and dissemination of brochures on noise, prepared by the Council on the Environment of New York City. Several noise brochures have been distributed nationwide to colleges, libraries, community groups, and individual citizens. These brochures include topics such as how noise affects health, learning, and quality of life; how to protect against noise, both personally and in the home; how to get help from public agencies; and how to organize citizen action groups.

Noise Awareness Conferences

In its attempt to educate a wider audience about the dangers of noise, the Council on the Environment has sponsored noise conferences at several colleges in New York City. These conferences have brought together a team of noise experts including audiologists, psychologists, environmentalists, public officials, and consumers. Invitations were extended to the community at large through college mailings and media advertisements. Noise brochures were distributed to attendees and they were urged to bring the message of the conference to their family members, friends, and associates. Attendance has generally been good, and responses from the attendees have been strongly positive.

Resources

Various organizations have developed anti-noise programs and materials for both school children and adults. They all believe that education is an important element in lowering the decibel level of a living environment that is becoming noisier at an ever-increasing rate. So that more people may be made aware of the dangers of noise as well as of ways to abate it, these organizations stand ready to assist anyone who is interested in pursuing an anti-noise campaign.

Further information can be obtained from the following organizations:

The Council on the Environment of New York City (CENYC)
51 Chambers Street, Room 225
New York, New York 10007
(212) 566-0990

The Big Screechers is a highly effective citizens' group in New York City that has led the increasingly successful fight against subway and elevated train noise.
c/o The Council on the Environment of New York City
51 Chambers Street, Room 225
New York, New York 10007
(212) 566-0990

The New York League for the Hard of Hearing is the oldest organization in the United States serving the needs of hearing-impaired persons.
71 West 23rd Street
New York, New York 10010
(212) 741-7650
TDD (212) 255-1932

Self Help for Hard of Hearing People, Inc. (SHHH) is active in chapters around the country as a support group for hearing-impaired persons and in the dissemination of preventive strategies.
7800 Wisconsin Avenue
Bethesda, Maryland 20814
(301) 657-2248
TDD (301) 657-2249

Hearing Education and Awareness for Rockers (HEAR), founded by rock musician Kathy Peck of San Francisco, is a foundation devoted to the prevention of hearing impairment in musicians and audiences from amplified music.
P.O. Box 460847
San Francisco, California 94146
Office: (415) 441-9081
24-hour Hotline: (415) 773-9590

The American Speech-Language-Hearing Association (ASHA) is the largest national professional organization for audiologists and speech-language pathologists, and serves as an international referral source for anti-noise activists and resources.
10801 Rockville Pike
Rockville, Maryland 20852
1-(800) 638-8255 (The Helpline)

REFERENCES

Arthur, D.A. 1988. Hearing conservation in an educational setting. In *Hearing Conservation in Industry, Schools, and the Military,* ed. D.M. Lipscomb. Austin: pro. ed.

Downs, M. 1990. Personal communication.

Florentine, M. 1990. Prevention Strategies: Education. In *National Institutes of Health* (NIH) *Consensus Development Conference on Noise and Hearing Loss. Program and Abstracts.*

FRAGER, A.M. 1986. Toward improved instruction in hearing health at the elementary school level. *Journal of School Health* 56:166-169.
FRAGER, A.M., and KHAN, A. 1988. How useful are elementary school health textbooks for teaching about hearing health and protection? *Language, Speech, and Hearing Services in Schools* 19:175-181.
KILLIP, D.C., LOVICK, S.R., GOLDMAN, L., and ALLENWORTH, D.D. 1987. Integrated School and Community programs. *Journal of School Health* 57:437-444.
LASS, N.J., WOODFORD, C.M., LUNDEEN, C., LUNDEEN, D.J., EVERLY-MYERS, D.S., MCGUIRE, K., MASON, D.S., PAKNIK, L., and PHILLIPS, R.P. 1987. A hearing-conservation program for a junior high school. *Hearing Journal* 40(11):32-40.
LEWIS, D.A. 1989. A hearing conservation program for high school-level students. *Hearing Journal* 42(3):19-24.
MELNICK, W. 1984. Evaluation of industrial hearing conservation programs: a review and analysis. *American Industrial Hygiene Association Journal* 45:459-467.
ROESER, R.J. 1980. Industrial hearing conservation programs in the high schools (protect the ear before the 12th year). *Ear and Hearing* 1(3):119-120.
SUTER, A.H. 1986. Hearing Conservation. In *Noise and Hearing Conservation Manual.* 4th ed., eds. E.H. Berger, W.D. Ward, J.C. Morrill, and L.H. Royster. Akron (Ohio): American Industrial Hygiene Association.

Subject Index